Toyota Methods and Operating Models

Toyota Methods and Operating Models

Achieve business success with the Toyota Way

Stefano Cortiglioni, Leonardo Salcerini and Danilo Verga

KoganPage

First published in Italian in 2017 as *Il successo continuo: L'eccellenza Toyota dalla via Emilia all'Europa* by Ulrico Hoepli Editore S.p.A.

First published in Great Britain and the United States in 2020 by Kogan Page Limited

2nd Floor, 45 Gee Street
London
EC1V 3RS
United Kingdom

122 W 27th St, 10th Floor
New York, NY 10001
USA

4737/23 Ansari Road
Daryaganj
New Delhi 110002
India

www.koganpage.com

ISBNs

Hardback 9781789663051
Paperback 9781789663044
Ebook 9781789663075

British Library Cataloguing-in-Publication Data

A CIP record for this book is available from the British Library.

Library of Congress Control Number

2020934437

Typeset by Integra Software Services, Pondicherry
Print production managed by Jellyfish
Printed and bound by CPI Group (UK) Ltd, Croydon CR0 4YY

Dedicated to my family, in particular to my wife Roberta and my kids Alberto, Lorenzo and Elisa, who supported me and gave me the strength to face this beautiful professional experience.

Stefano Cortiglioni

To Gigliola, Jacopo and Chiara, who give a purpose to my life, and to all my family and friends for their constant support and motivation.

Danilo Verga

CONTENTS

INTRODUCTION BY THE ITALIAN TOYOTA MATERIAL HANDLING ORGANIZATIONS

by Danilo Verga[1]

It was a great pleasure for me to accept the invitation I received from the management of the Italian Toyota Material Handling organizations (Toyota Material Handling Manufacturing Italy and Toyota Material Handling Italy), which asked me to join forces with the same winning, proactive management team to create an authoritative, ambitious and challenging publication.

The strategic goals of this initiative are the analysis of Toyota values and the application of the principles of the Toyota Way and Toyota Production System provisions to Italian manufacturing sites and commercial network. This publication will specifically examine the strategic choices that were taken and the good business practices enacted by the same management team, which has been capable of achieving constant success in the transformation of the entrepreneurial effort formerly known as CESAB S.p.A.

The results of such in-depth studies and successful performances, which have spawned and qualified new business management excellences along the via Emilia, constitute a notable business and managerial reference material. The management of the Italian Toyota Material Handling organizations have developed this publication with the aim of sharing such assets and repaying their territory and community.

Therefore, this publication aims to define and illustrate the application of Toyota's principles and transformation processes that were developed by the Toyota Material Handling organizations operating in Italy, specifically for the Italian manufacturing and commercial environment. Particular attention shall be given to the peculiar conditions that had to be faced, to the issues that were resolved, to the challenges that were faced during the process of transforming a local, family-run enterprise into a modern, innovative and successful international organization.

The unique content that is shared on these pages constitutes a valuable in-depth analysis, offering a solid set of guidelines that were successfully implemented to achieve:

- organizational change;
- corporate transformation;
- alignment of the Italian manufacturing, commercial and service organizations to the new challenges of the current industrial revolution.

Moreover, this data represents a reference model that can be followed to repeat our effective institution of a new, cooperative form of company management that focuses on innovation and growth.

Several esteemed researchers and a large number of successful consultants have written abundantly on subjects such as the Toyota Way values, the principles of Lean Production, Lean Leadership, Total Quality Management, World Class Manufacturing and the Toyota Production System, commonly referred to as TPS. This publication is being developed by Toyota managers wishing to describe the challenges they had to face first-hand, the original vision they envisaged and the Toyota strategies they implemented for Toyota's Italian factories, explaining how they succeeded in the creation of an ecosystem of suppliers and customers, both direct and indirect, surrounding the Italian Toyota Material Handling organizations.

Finally, with this new publication the management team wishes to describe and share:

- the difficulties faced during the corporate transformation process, and how they were overcome;
- the operational methods that were designed and implemented;
- the organizational innovations that were developed for processes and products;
- the cultural changes that were pursued and realized;
- the processes that were implemented to improve operations;
- the evolution and the outcome: the targets of excellence and continuous success that were planned and achieved.

Understanding 'The Toyota Way' and 'Beyond the Crisis'

This book, created by the Italian Toyota Material Handling organizations, centres around two themes: The Toyota Way and Beyond the Crisis. These are two conceptual frameworks that will characterize the publication. Here I will briefly explore such elements, offering a defined structure of the perspective and topics regarding this new and unique publication.

The Toyota Way

First of all, the Toyota Way. The Toyota Way, as it is widely known, is the set of values that has driven the Italian managers of Toyota Material Handling while developing and expanding the Toyota excellences located in Italy, specifically along the via Emilia. The Toyota Way includes all the values that define the vision and contribute to determine and structure the strategic and organizational mission of the company. The Toyota Way is built upon the five pillars that define a system of personal development, itself created upon the values of teamwork and respect, and a continuous improvement system that is defined by the principles of accepting new challenges, constant improvement (*Kaizen*) and direct and tangible orientation of the source of real operative scenarios (*Genchi Genbutsu*).[2]

Beyond the Crisis

'Beyond the Crisis' is the second critical element of the publication.

It encompasses all the dynamic, evolutionary values and dispositions that were part of the process that brought constant innovation, growth and improvement, identifying the **direction, innovation landscape and transformative paradigm of the company**, founded on the values of respect and development, on responsible and structured cooperation, on teamwork and on constant innovation.

Beyond the Crisis: the direction

The idea is to illustrate how the company, in an environment characterized by constraints and issues caused by the market crisis, implemented dedicated **corporate guidance** and **temporal frameworks** to drive the business towards growth and expansion.

The framework includes an analysis of the corporate events (all of them based on the values of the Toyota Way) that occurred during the observation period: the main, defining moments of transformation of the original local business, from 2000 until 2016. Starting from the restricted landscape of a local business environment, the authors of this book, acting as the Italian managers of Toyota Material Handling, implemented a series of activities that looked into the future, predicted the evolution process of the fourth industrial revolution and acknowledged the technological advancement of artificial intelligence (AI) and its application in various sectors, a process that will eventually take us into the fifth industrial revolution.[3]

Beyond the Crisis represents a qualified framework, taking into consideration the transformative managerial actions that were taken over a long period of time, characterized by an extended crisis that affected economy, finance and markets. Indeed, the timeframe subject of this analysis begins in 2000, well before the crisis began and deepened (2008), includes the peak crisis period and stretches until 2016, with an overview of future developments.

Beyond the Crisis tracks the period of time during which the original company, BT CESAB S.p.A., was transformed into today's organization, the Italian Toyota Material Handling business. This is the timeframe during which Toyota's Italian management implemented, in a peculiar and highly distinctive way, the values of the Toyota Way and the principles of the Toyota Production System. These values and principles were the basis upon which to create a new corporate vision, which in turn drove our strategic thinking and defined the long-term strategy for corporate innovation, including the transformation of production and commercial processes.

This strategy has determined how to develop and restructure the factories and their manufacturing systems, how to set up the whole new commercial approach, how to introduce new services and develop corporate culture.

In short, these 'direction guidelines' focus on a qualified, structured and well-oriented transformation which, once the values of the Toyota Way are embedded, can contribute to outline a new, distinctive innovation landscape for the company.

Beyond the Crisis: the innovation landscape

The new corporate culture adopted by the Italian organizations inside Toyota Material Handling is focused on a landscape of innovation. The company develops its paradigms around the predominance of **human value** and on a renewed **centrality of the individual**. By human values, we include the company evolution and changes, the interactions with suppliers and stakeholders, and the relationships with qualified and selected suppliers upstream, dealers and direct customers downstream.

Beyond the Crisis: the innovation landscape of the company and the architecture of the new cooperative management for 'strategic and transformative buildership'[4]

The new innovation landscape of the company is focused on human values and on the newfound predominance of the people, actively driving evolution, alongside the corporate changes required for the transformation process, implemented with the application of the principles of the Toyota Way and the adoption of new digital (and non-digital) technologies in a considered and well-balanced manner. This new innovation landscape satisfies:

- the new needs that were expressed by the workforce, wishing to contribute directly, to be more involved and to add value, which produced a new strategy that is in fact a new way to manage the company. It is a 'New How', capable of channelling the energy of each worker, integrating and connecting it with all the others in order to create value through synergy;
- a growing request by the workers to be more involved and to 'count more'. This implies giving more value to each worker, to each person, creating a new, increasingly evolutional, dynamic and shared leadership. A new environment with a new constructive and regenerative function, which we will call with a neologism, 'buildership': a new flow, a new concept of values and behaviour, a fabric woven with motivation and infused with **new constructive energy**. Buildership means giving value to people, it means a shared leadership that is regenerative, motivational, responsible, open, capable of creating value for the people and for the single worker, for the organization, for the company and, exponentially, for the community and the surrounding territory.

Buildership's conceptual architecture reinterprets and supplies a valid answer not only to the need to put people first, but also to the increasing need to share corporate knowledge, redefining the involvement of people into company activities and the common good of the business. Buildership is therefore the 'new corporate paradigm', which reinterprets and enriches the common concept of leadership, while underlining the need for a new brand of adaptive leadership, capable of understanding the surrounding evolution and changes and to adapt accordingly. Moreover, buildership not only can adapt, but it can also forecast and prevent changes, reconfiguring the vision of the company, which is in turn a source of strategic value regeneration. To this effect, buildership also includes a pre-emptive cognitive side, capable of forecasting and interpreting new trends to the point of becoming a trend-setter, bringing changes and innovation to the company's strategic vision.

Buildership can ensure a systematic, continuous and multimodal alignment of the company organization, so that it is always tuned and ready to face future changes. This is because it can create flexible organizational plans which, while they maintain the positive aspects of the key elements that are common in a traditional hierarchical organization. They can also develop a new organizational approach, evolved and ready to be implemented, typical of a non-hierarchical organization capable of dealing with the widespread transformation of organizational duties, which are becoming extensive and based on sharing.

It means spreading and sharing responsibilities, where each person is accountable for their involvement and production volume and for helping productivity of other people, working both within and outside the company organization.

Such definition and reinterpretation of roles, vision and participation of the worker to the constant growth of the company can regenerate value by the workers themselves, who feel more involved in the process of buildership through motivation and inclusion, making them become an active part of the company's organization: they are an integral, proactive, defining element of the organization itself. Such an environment can deliver and empower; it can show how synergies are fundamental, and it can underline the importance of the idea of extended performance, as a group and as an individual.

Beyond the Crisis: a transformative paradigm

The concept of 'Beyond the Crisis' includes the new **transformative paradigm** (dynamic, positive and evolved) of corporate choices and policies; it is a paradigm in constant evolution, characterized by the aspects that qualify and determine the corporate change processes, by anticipation, amplification and acceleration of the same change processes, and by other important elements (such as adaptation, flexibility and autonomation[5]) that create connection, contamination, sharing and continuity of operations.

Focuses and goals of the publication by the Italian Toyota Material Handling organizations

This publication focuses on the description of the corporate strategic and managerial choices that were taken, of the operative changes that were implemented, of the challenges that were faced and won, illustrated first hand by the Italian management of Toyota Material Handling.

Specifically, the contents of this book will explain in detail the complex processes of analysis, planning, execution, control, implementation and diffusion of the company initiatives that were applied by management, targeting:

- the strategic, operational and innovative activities that were developed for the transformation of the manufacturing and commercial processes;
- the educational and consulting initiatives that were planned and developed by the management and were offered to other businesses, schools, organizations, universities and business schools during the last few years.

Toyota methods and operating models

The application of Toyota's principles and values has created manufacturing and commercial excellences, crucial to the continuous success of the original Emilia plant in the period ranging from 2000 until 2016. The old business has evolved into a modern, international manufacturing company and a successful and complex commercial organization.

Toyota Methods and Operating Models introduces the challenges that were accepted and faced by the managing team of the Italian Toyota Material Handling, the good practices that were developed and integrated with specific processes, organizational guidelines and application developments, resorting to Toyota knowledge, culture, methods, systems, operations and its advanced labour system.

Such a universally known organization, once integrated with methods, values and corporate management systems, has brought continuous success in all business sectors; its unique people-centric approach has determined the strategies that had to be implemented, while defining how to perform several activities dedicated to training and expanding workforce skills and teamwork capabilities.

Notes

1 Author and coordinator of the Toyota Way series.

2 See the second part of this book.

3 The development and implementation of digital technologies, a tangible activator of social transformation, requires the company to adapt its corporate organization, production processes and commercial strategies, regardless whether the offer is for goods, services or solutions. It creates the need for deep, urgent and unprecedented organizational transformations, so that the company can maintain its efficiency levels. An example of this transformation is the revolution brought by the development of online sales by primary businesses (such as Amazon), which has worsened the crisis of several companies operating in the retail sector such as bookshops, supermarkets and shopping centres, causing closures and job losses.

4 The concept behind a strategic and transformative buildership comes from: the results of 30 years of research, investigation and consulting interactions developed with major businesses, organizations, entrepreneurs and managers of international concerns by the author of this introduction; a specific research effort by the same author on the development of innovative management of 'cooperative clusters', on alliances and business networks and their cooperative management. Strategic and transformative buildership includes all conceptual innovations related to the transformation of open organizations, with the right understanding of the value of putting the person in the core of operations, and seeing technological transformation as an essential platform for their organization and a strategic element for development. Strategic and transformative buildership is what defines the development of the new business production systems and of market management, and includes the strategic factors deriving from the evolved applications of the fourth industrial revolution (Industry 4.0). It also includes the values, the vectors and the evolutional and transformational advantages brought by the early adoption of technological advancements and of the game-changing innovations in all the sectors involved with artificial intelligence, which is in itself at the core of the upcoming fifth industrial revolution.

Finally, the buildership that goes beyond the current boundaries of adaptive leadership is the cornerstone of the new management system, described as 'New Cooperative, Energetic, Responsible, Innovative, Open, Sustainable, International and Integral Management'.

5 Autonomation or '*Jidoka*': detecting and identifying anomalies. A principle stating that any quality issues must be investigated by stopping the production process, reducing waste to a minimum. This concept will be detailed at a later stage.

PREFACE

by Ambrogio Bollini[1]

It was with great pleasure, and with some excitement, that I accepted to write the preface for this book which, although focusing on technical matters, does not ignore the human experience that accompanies every process of change.

Toyota's production system, commonly called TPS, was designed and adopted in several sites, with generally positive results. However, only the direct involvement of the people, bringing their humanity and willingness to face the challenges ahead, allowed the results to become excellent, both in terms of manufacturing system and corporate growth.

The story told in these pages shows how a different way of doing things, if fully understood and implemented, can affect the long-term outlook of companies and people.

Technically speaking, the concepts that are typical of Lean Production have long been absorbed by our industrial culture; TPS, however, is something different, something that goes further.

Nowadays we are accustomed to rely on diagrams and charts of all sorts to illustrate techniques that might reach a fairly high level of complexity; what we struggle with is to explain simple concepts, and TPS has a clear and extremely effective concept.

Personally, I like to describe it as the 'structured application of daily common sense'.

Is it too simple? Perhaps, but its strength lays in a simple, clear method that sees facts as facts and opinions as opinions, one that puts the factory and its workers at the core (*Monozukuri*).

I am convinced that the best way to really understand the meaning of TPS is to observe it, to see how it became the core of the companies that have adopted it: the aim should not be to learn its basic techniques faster, but to understand the shift in paradigm that this system requires at the heart of the company's organization. It forces companies to overcome taboos and to state the centrality of the individuals, each one of them taking part and contributing to corporate success with intelligence, humanity and fallibility.

If one doesn't believe in this unquestionable truth, the application of the same method in different contexts will deliver radically different results. What really matters, for this system to work, is people.

Personally, I had the privilege to be part of this unusual and exciting *new world* since 2001. My memory goes back to my first few days at Toyota, when I had the honour to meet an elderly *sensei*, and I naively asked him what were the pillars of Toyota's production system, expecting to be given an erudite description of TPS techniques.

The answer I received instead was as simple as bewildering; in a nutshell, I was told that the secret was 'being able to do a lot with a little'.

These words have a deeper meaning. They tell the story of a national community that with very limited resources had the creativity and motivation needed to re-emerge from the grim aftermath of a devastating war: doing things the simple way, with method and discipline, without massive investments, placing at the core of everything the factory and the intelligence of people, respecting individuals, society and environment. A system called *Monozukuri*.

Having said that, it is obviously crucial to understand and adopt the techniques and methods that are the backbone of TPS and these are clearly illustrated in this book. However, all these changes can only achieve limited success if they are not implemented in a context where human resources are firmly at the core of the process, or if there is no real and tangible motivation to change the way things are done, if the idea of a drastic operational shift is not appealing enough.

The operational side of an enterprise is unsurprisingly the most receptive and ready for change. However, the main role is played by management, in charge of implementing TPS and operating according to its rules, guiding the whole company transformation process.

Last but not least, it is necessary to take into consideration the concept of quality, and the kind of culture that is capable of achieving it.

The term 'quality' has been the subject of several different interpretations over time, almost exclusively related to the product.

This is not enough. TPS needs quality to be at the centre of all the company processes, along the whole supply chain and at every step during the creation of value. This allows the production of only what is needed, when it is needed, at a price that the customer is willing to pay, while respecting people and the environment.

It is something that's understandably easier said than done; again, people are what makes the difference, and the most successful enterprises prove this.

The question is whether the *Monozukuri* model will keep on being effective, or if it will need to be reinterpreted dynamically, in order to adapt it to an ever-changing economic context. Whatever happens, it will be even more critical to rely on creativity, on talent and on the added value that every person can bring; it's people, with their strengths and weaknesses, who must always be placed at the core of everything, which is what *Hitozukuri* actually means.

Enjoy the read.

Note

1 Senior Vice President Supply, Toyota Material Handling Europe, Former General Manager and CEO of Toyota Material Handling Manufacturing Italy S.p.A., President and CEO of LTE Lifttruck Equipment S.p.A., President of Simai S.p.A.

FOREWORD TO THE INTERNATIONAL EDITION

by Stefano Cortiglioni

When I was asked to write a brief introduction to the English edition of the book, highlighting the elements that are relevant in any cultural context, I immediately thought of the transformation of the Toyota group's Swedish and American plants, which I experienced indirectly. However, the Italian case study described in this book is universally applicable.

The first edition of this book was published in Italian with the title *Toyota Way: il successo continuo*. The text describes the transformation of two Italian businesses after they were bought by the Toyota group.

Firstly, the production plant was transformed into an international manufacturer of counterbalanced forklifts for the Toyota brand, secondly the Italian sales network merged into a single Toyota sales network.

Common to both of these transformations was the adoption and application of the Toyota managerial style, where the skills of the workers are preserved and maintained as they are of major importance during a period of change.

This method has been applied by Toyota around the world, resulting in many successful business transformations. The model described in this book is based on solid methodology, successful styles of management and leadership and deep business values which have a universal application.

INTRODUCTION

by Stefano Cortiglioni, Leonardo Salcerini and Danilo Verga

The book *Toyota Methods and Operating Models* is divided into three parts. The first two focus on the productive and commercial aspects, while the third part includes the data of the Appendices and further information regarding TMHIT (Toyota Material Handling Italy). The valued contribution by Ambrogio Bollini introduces the spirit driving the content of this book: giving the overview of an integrated system composed of values, entities, occurrences, perspectives, solutions and good practices, all together constituting a unique effort towards innovation. This is the story of the successful transformation of a business that was originally family-run and focused on the local market into a modern, international manufacturing company.

This success case, illustrated by the managers of the Italian Toyota Material Handling organizations themselves, is first and foremost a benchmark for comparison, an indispensable resource of methods and operations for manufacturing companies. This business case represents a new 'corporate and managerial template', a practical system that defines the guidelines to transform radically the organization of small and medium companies in the manufacturing sector (but not necessarily from that industry segment), both in Italy and in Europe, and for their commercial organizations.

Moreover, the contents of the two main parts of the book, together with the substantial Appendix B, tell a detailed and fascinating story about the successful transformation of the enterprise into a multinational company.

First of all, such integrated business transformation arises from a systemic and extended vision of the business: an amplified, all-encompassing vision that goes beyond the traditional boundaries that characterize the company, involving and affecting all the cooperative processes of the businesses upstream (the suppliers) and downstream (dealerships and customers). This augmented, integrated vision of the business, which encompasses values, priorities and boundaries, becomes a **business/community**, chosen and prioritized over the traditional, functional vision, which is nothing but the independent collection of different business functions.

This book illustrates the transformation of a traditional, medium-sized Italian family-run manufacturing business into a modern and successful international enterprise, structured into the Italian organizations of Toyota Material Handling.

CESAB S.p.A. is a company founded in Bologna in 1942, operating in the production and sale of forklifts, which was acquired in April 2000 by BT Industries AB; a few months later, in July of the same year, Toyota Industries Corporation (TICO) bought BT Industries. In 2001, Toyota Industries Corporation became the world leader in the materials handling market; in 2005 the company created a new worldwide organization, Toyota Material Handling Group (TMHG), structured in Europe under the name of Toyota Material Handling Europe (TMHE).

In 2006, Toyota celebrated 50 years of activity in material handling, while Toyota Material Handling Europe began an integration process for the commercialization of Toyota and BT CESAB branded forklifts.

The success case we are going to study shows the continuum of choices and corporate activities that drove the gradual transformation of CESAB S.p.A. into what today is known as the organization of the Italian companies of Toyota Material Handling. As already mentioned earlier, the first part of the book describes the evolution stage of the production processes and corporate operations, including the innovations that were implemented in the organizational processes of the network of suppliers. The second part introduces and explains the evolution of its commercial strategies and distribution network. The historic meeting of October 2009, when the two indirect sales networks of the company were merged, is the core of the complex process of commercial and organizational change, which gave birth to a new concept of cooperative management.

These two parts of the book constitute a unique managerial and strategic approach; they focus on the challenges that had to be faced, illustrating the managerial and organizational developments that characterized the continuous success and the growth of corporate results.

The book illustrates the managerial activities that were carried out and the operational choices that were made, exploring the various viewpoints and the processes that range from development to sales.

The strategies and operational choices here explained were planned and implemented by the Italian managers of the Toyota Material Handling organizations, who have combined their efforts to successfully transform a local manufacturing organization into a market leader with international standing, through the appropriate application of the principles of Lean

Leadership, of Toyota Way's values and of the related Toyota production management systems (ie TPS, Toyota Production System), while respecting the centrality of people and territory.

The parts dedicated to the contributions directly given by the main figures of this transformation are focused on strategic choices, company reorganization and corporate operational activities. With their decisions and activities, the Italian managers of Toyota Material Handling contributed to qualify and adapt, with Italian creativity and wit, the application of precepts, values, dispositions and principles of the Toyota Way and TPS, creating unique operational environments, in line with the reference market and strongly oriented towards the centrality of human values.

The picture that is depicted here shows a company and management that are strongly innovative and focused, far from the dictates of Fordism and consistent with the new cooperative management of an open, human and successful business.

The analysis of the choices implemented by the Italian management of Toyota Material Handling, and the results that were subsequently achieved, represent a complete overview of the methods that were implemented, making this book a tool that can be a valuable support for those who wish to achieve the same results.

The managerial choices that are illustrated within these pages are the result of a unique interpretation of TPS's principles and the values of the Toyota Way, contextualized and adapted to fit the context of the Italian forklift market.

The original aspects of Italian culture intertwined with the Japanese approach to things; mutual respect and great curiosity created a blend, each contributing with their best features and characteristics. The companies achieved total integration because the management decided not to copy, but to acknowledge and adapt; driven by the curiosity to understand the underlying values, it did not stop at superficial and prejudicial assessments. This decision turned out to be the winning choice.

The exploration of the choices and activities at the core of the methods and actions that were implemented can be a valuable resource for small and medium-sized manufacturing and commercial companies wishing to focus on transformation and growth.

Far-sightedness and a system-wide respect for all those who cooperate with the company made it possible to maintain continuous market success.

Constant performance and production growth, goals achieved through communication and the active involvement of people, have established a strategic and indispensable internal cooperation, expanding from the business to the suppliers of the district.

Moreover, management actions have affected the transformation of corporate culture and of the system of competences inside the company, fostering human capital and ultimately turning the business into a constantly improving and growing concern, a true and successful 'learning organization'.

The growth of the company, helped by teamwork and the professional development of all the individuals involved, was the obvious result of the deep transformation of production and factory processes, a transformation promoted and driven by carefully listening to the market and its customers.

In turn, this has triggered the innovation process in manufacturing and commercial activities, and a gradual and system-wide attitude of constant attention and prioritization of the customer, which created differential value and mutual satisfaction.

This new corporate vision has deeply affected manufacturing and commercial cultures, bringing constant improvement and a renewed positive tension towards the development of a widespread entrepreneurship which looks into the future while being firmly rooted in the local territories for quality, beauty, increasing productivity and far-sighted practicality.

Who is this book for?

The contents of this book are aimed at entrepreneurs, managers, consultants, researchers and experts in business transformation. It is useful for entrepreneurs and managers because it helps to understand what is really indispensable for the company, indicating the crucial factors for the transformation of manufacturing and commercial processes.

All the content, including case histories, methods and activities, is focused on real-life situations, on elements that are tangible and capable of creating added value. The managers of Toyota Material Handling Italian organizations, authors and drivers of the change and transformation processes of the company, wish to help the reader understand what is really useful and indispensable for the correct management of the business, privileging new inside and outside cooperative harmonies, following synergies that can create value for growth and the transformation of the production and commercial processes of a modern business.

Part One
From CESAB to Toyota Material Handling Manufacturing Italy

by Stefano Cortiglioni

Introduction

I often wondered whether it made any sense to write another book about Toyota, and what I could personally add to the existing bibliography. It is true that a lot has already been written, but I also believe that my personal experience can make a difference and add value: working for more than 10 years for the highest performing company in the world allowed me to understand its systems and learn the details and secrets that are lacking in other books and publications.

My experience with Lean Production, which from now on I will call by its proper name, Toyota Production System[1] (TPS), began with Ducati Motor Holding. During that period, I had the chance to discover the main aspects of TPS, together with the privilege of seeing it live in action. I was a young engineer back then, and I was immediately drawn and fascinated by that production model, so I tried to examine each of its aspects in detail.

This caused my first contacts with 5S, *Kaizen* and Just In Time: the capability to implement them and to fit them into daily practices made their great potential increasingly stand out each day.

The years spent with Ducati were filled with projects and implementations, and were professionally very satisfying. Working in the automotive sector for a company manufacturing limited volumes (about 200 motorcycles per day) of remarkable products with high seasonality, pushed us towards the rationalization of the manufacturing system, adopting innovative strategies such as the *one piece flow* and the *Pull*, which will be discussed in detail in the next chapters.

My real breakthrough with the understanding of TPS happened in 2005, when I accepted the offer to join the Toyota group, fascinated by the opportunity to learn all the elements that I had studied and applied with Ducati directly from the source.

This opportunity came from CESAB S.p.A, now Toyota Material Handling Manufacturing Italy (TMHMI), a company based in Bologna, specialized in the development and production of counterbalanced forklifts, a company that had just become part of the Toyota group and was switching from local ownership to become part of an international conglomerate.

A few months after I started working for them, the Japanese parent company created a project for the education of European trainers, which would promote the diffusion and application of TPS inside the newly acquired production sites. When this opportunity was mentioned to me, I did not waste a minute and accepted it right away. I was therefore sent away for training, first to the group's manufacturing units in the United States, then directly to the parent company in Japan.

My visit to the Toyota factories allowed me a first glance into that magic world, where everything seemed planned down to the tiniest detail, and the idea of something unexpected happening was not contemplated. During my time spent with the company I had the chance to travel to Japan several times, visiting most of the group's production sites, each time enhancing my knowledge with further information about how they organized and carried out their work.

I still remember with some excitement an educational event organized by a Toyota senior advisor, which required for me to spend some time actively working on a forklift assembly line inside one of the company's plants in Japan. Since I was familiar with acquiring knowledge through theory rather than practice, I must confess that I was quite sceptical about the actual point of spending 15 days wearing safety shoes, holding a stopwatch and a notepad in front of an assembly line, for 8 hours a day, noticing potential *Kaizen* (improvements) and recording the sequence of operations while checking the worker's motions. In hindsight, this was probably the best possible training I could receive. Living the assembly line alongside the operator (Toyota calls it *Genchi Genbutsu*) turned out to be a fantastic experience: I felt immersed in a huge clock, of which I could spot the mechanisms and each of their functions, and I could finally understand the complexity and precision of its structure. The line was assembling forklifts with a *takt time* (TT) of about 4 minutes. Next to the worker there were two ropes, a yellow and a red one. If the operator pulled the yellow rope, the team leader (TL) would appear; the team leader's job is to assist the workers along the assembly line. By pulling the rope, you could hear an audio signal and the *Andon* showed the station that needed assistance. Conversely, the red rope would stop the line, fully in compliance with the concept of *Jidoka*, which means 'to bring anomalies to light'. When the worker realized that they could not respect the assembly time, they would call the TL, who would intervene immediately. I still remember very well how the TL, without communicating with the worker, was able to understand the issue and start helping with the assembly so that it could be finished on time. If the issue could not be solved within the TT, the red rope would be pulled to stop production. The *Andon* would show the station that suffered the issue, and the TL would rush there to try and solve the problem.

Next to the station, the parts were placed following a precise layout that allowed the worker to move as little as possible; most of the parts were placed into plastic boxes, carried to the station by roller conveyors. A series of 'small trains' and AGVs (automated guided vehicles) were supplying the parts, carrying them from the main warehouse straight to the lines.

Each part was meant to be used for a specific order, and the worker had all the elements needed for its assembly right by their station.

The logistics department's duty was to retrieve the parts required for the assembly of each and every forklift from the supermarkets; the only task of the operator was to assemble them correctly, trying to minimize their motions.

Once I returned to Italy, I had the task of transforming a plant, which until then had worked following traditional production systems, into a Toyota factory. I had to mediate cultural and organizational differences and acted to implement a huge shift in corporate culture.

Note

1 The term Lean Production (Womack, 1991) was coined in a study conducted on the production methods of the main car manufacturers compared with Toyota's. The study revealed that Toyota was by far superior. The results of the study were published in *The Machine that Changed the World: The Story of Lean Production*. Toyota's system was later studied and copied by the industrial world; many terms were used to emancipate the system from the name Toyota, but the philosophy and the methods derive from Toyota Production System.

01
The Toyota Group

Toyota was founded in 1918 in Japan, originally named Toyoda Spinning and Weaving Company. Sakichi Toyoda, its founder, was an inventor and innovator working in the textiles sector: he refined several technical aspects of the looms of the factory, bringing increased efficiency and better quality. His solutions saw the creation of a concept which is still one of the pillars of TPS: *Jidoka*, the autonomation. One of the main inventions of Toyoda is a weaving loom, called Type G, which stopped automatically if any of the threads of the fabric broke during operations. This principle caused two substantial revolutions:

- The machine guarantees its own quality, because it stops when it detects quality issues.
- The worker can follow several machines at once, because they must no longer review the quality of the finished product.

In 1926 Toyota Industries Corporation (TICO) was founded, while in 1936, thanks to the initiative of the founder's son, Kiichiro Toyoda (Figure 1.1), it was the turn of the Toyota Motor Company, heralding the expansion of the group into the automotive sector. Kiichiro Toyoda is the person responsible for the development of the second pillar of TPS: the concept of Just In Time.

All the tools that constitute what is called the Toyota Production System are designed on the basis of these two pillars.

TPS was first introduced in 1953, under Eiji Toyoda's presidency, father of today's TICO chairman, Tetsuro Toyoda; Taiichi Ohno was one of the architects and promoters of the idea. The need to expand their vehicles market outside Japan and enter the USA market meant that Toyota had to compete with the major carmakers at the time, including GM, Ford and Chrysler. Post-war Japan had very scarce economic resources, way behind those at the disposal of the huge American companies that dominated the scene; Toyota was forced to study alternative production approaches, systems that did not

Figure 1.1 The pillars of TPS

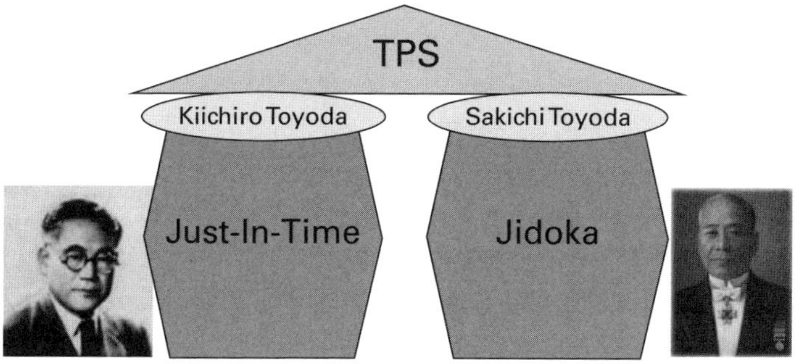

require massive investments and allowed the production of a wide enough range to satisfy the needs of several types of customers. While bigger companies could afford to build factories specifically for a single car platform, Toyota worked on flexibility and setup time reductions, so that the same production chain could be used for several models. Worthy of notice is the implementation of SMED[1] (Single Minute Exchange of Die) on bodywork presses, which allowed for very fast mould switches and substantially reduced the time required to change production from one car model to another.

Later on, Toyoda changed its name to Toyota, officially to end the association between company and founder; however, there is another unofficial motive, linked to tradition and superstition: the name 'Toyota' needs eight brushes to be written, a lucky number according to Japanese culture.

Today, Toyota Group (Figure 1.2) comprises 17 different companies, linked together by shareholdings and covering different business sectors.

Toyota Industries Corporation (TICO), aside from being the original company of the group, is active in four different business sectors, including the textile industry, where everything started (Figure 1.3).

Toyota Material Handling Group (TMHG) is one of the five units that are part of Toyota Industrial Corporation (TICO), a company headquartered in Nagoya, Japan, employing over 50,000 people and posting yearly revenues of over €16 billion. TMHG's business area is the development, manufacturing and sales of material handling solutions; the company is the world leader in terms of market share.

In addition to development, manufacturing and sales of counterbalanced and warehouse forklifts, the company offers turnkey projects for the creation of automated warehouses and logistics solutions based on the clients' specific needs. The group has several production sites, covering the main

Figure 1.2 Toyota Group in 2016

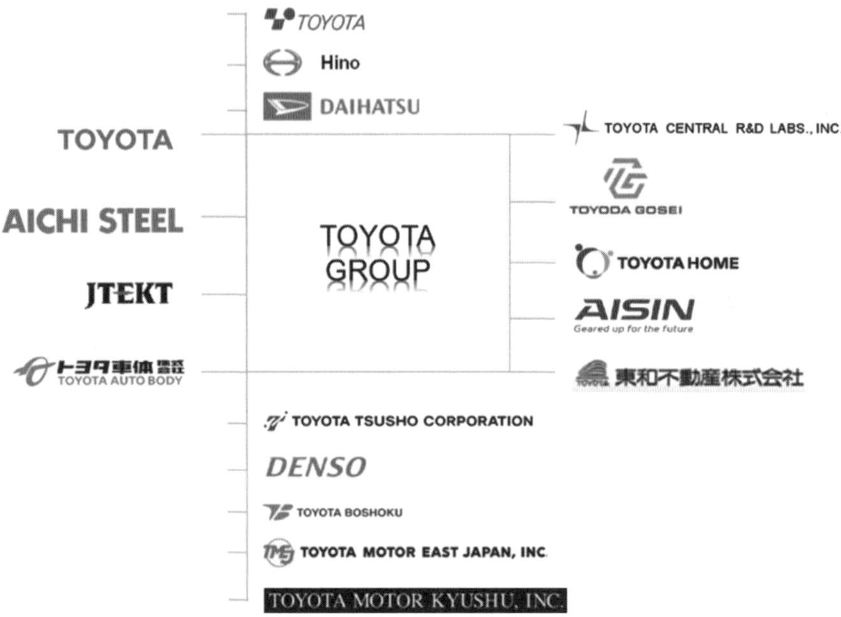

Figure 1.3 TICO business results in 2016

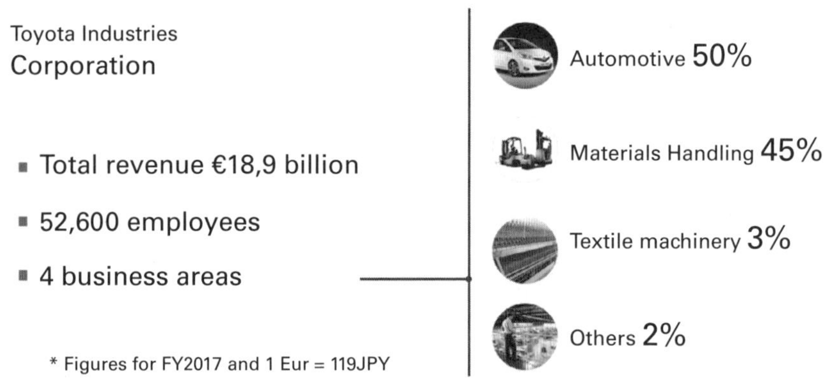

global geographic areas. Toyota Material Handling Europe (TMHE) is the European branch of TMHG (Figure 1.4), with the job of developing, manufacturing and selling products in the European markets (Figure 1.5).

The products are mainly destined for the European market and a few other locations; they are developed and manufactured in four production

Figure 1.4 Location of TMHG factories in the world

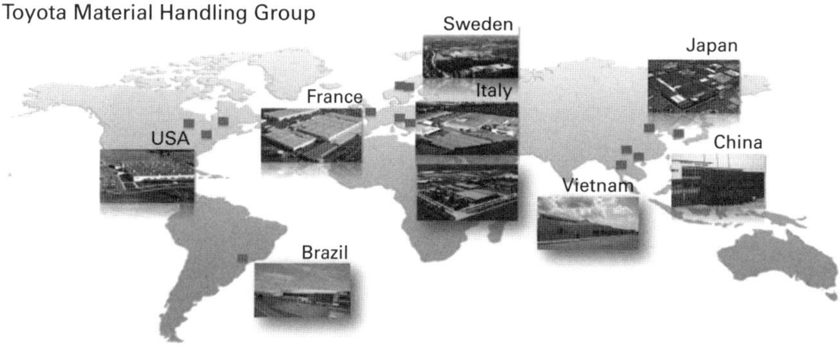

Figure 1.5 The data relating to TMHE

TMHE'S NUMBERS	UNITS SOLD IN 2016	REVENUE	EMPLOYEES	EUROPEAN COUNTRIES COVERED	OPERATIONAL FORKLIFTS
	85.000	2.0 miliardi	9.500	30	330.000

sites, all of them located in Europe (Figure 1.6). Each site is dedicated to a product type:

- TMHMS (Toyota Material Handling Manufacturing Sweden) deals with the development and manufacturing of the whole range of products dedicated to indoor material handling.
- TMHMI (Toyota Material Handling Manufacturing Italy) is focused on the development and manufacturing of the entire range of electric counterbalanced forklifts.
- TMHMF (Toyota Material Handling Manufacturing France) develops and produces the range of thermal combustion counterbalanced forklifts.
- LTE (Lift Truck Equipment) is the site in charge of development and manufacturing of masts for the entire group.

Figure 1.6 TMHE production and R&D European sites

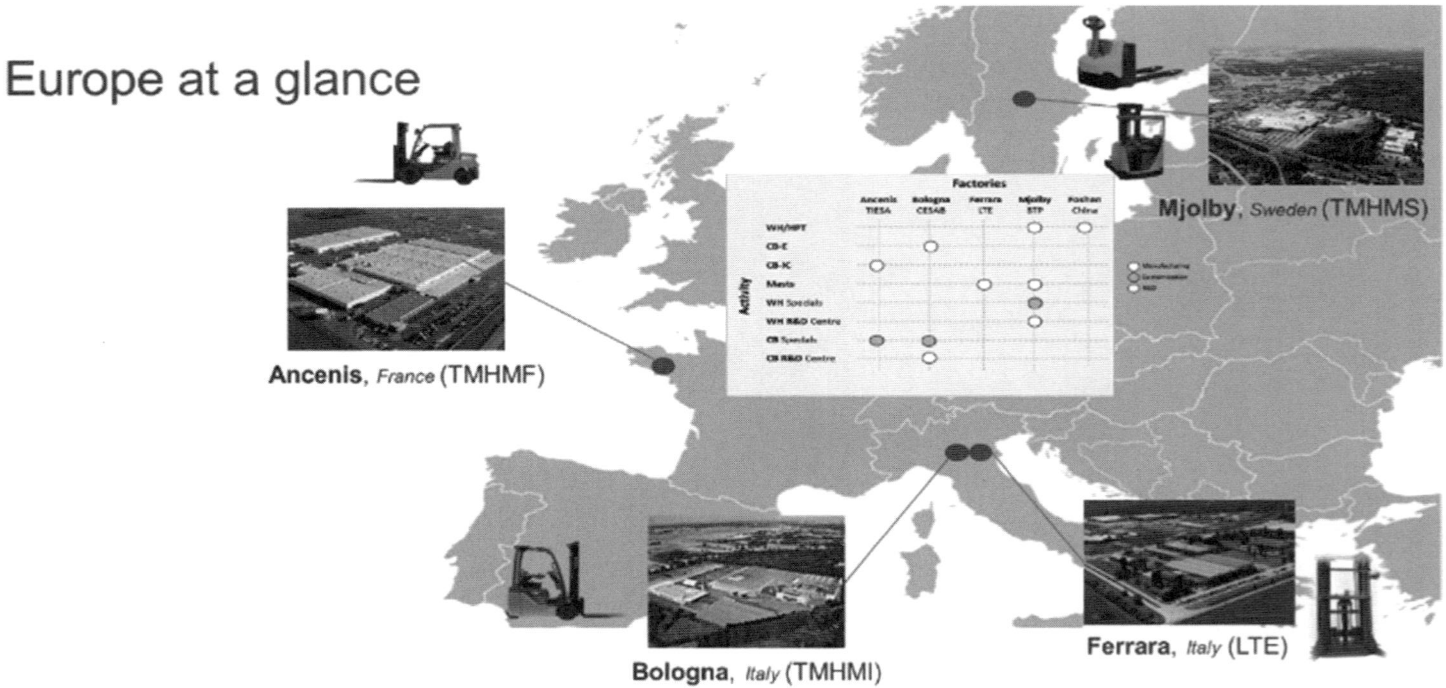

Note

1 SMED, Single Minute Exchange of Die (Shigeo Shingo, 1985), is a method applied by TPS. It was first used by Mazda in 1950 before being implemented by Toyota. Its aim is to optimize setup times and reduce them to a minimum. 'Single minute' means that the setup must take less than 10 minutes (single digit). The main procedure is to assess the operations needed for the setup, splitting them into those that can be carried out while the machines are working, and those that require them to be stopped. Each operation must be labelled, splitting those that must be performed when the machine is turned off from those that can be done when it is in motion. Each operation, in particular those to be done during downtime, must be reduced and optimized, so that they can be carried out in the shortest possible time.

02
CESAB's history

Before we begin illustrating the course of actions that transformed a classic Italian proprietorial enterprise into a TPS reference company within the Toyota group, it is useful to describe its history and products. CESAB, Carrellificio Emiliano Società per Azioni Bologna, was founded in 1942 by a Bolognese family with the scope of developing and manufacturing counterbalanced forklifts. The company achieved considerable success over the years, becoming the main Italian producer of (mostly electric powered) counterbalanced forklifts. The 1990s saw the company expanding into foreign markets. At the time, production was limited to a few thousand units a year, while the assembly organization resembled more a craftsmanship concern, rather than a production line. The long-term outlook that underlined the need to find an international partner pushed the management to make further investments in electric forklift technology; Pininfarina was involved to add value through Italian design. This period, which was very business-oriented, following what was in fact a pioneering stage, saw important strategic decisions taking place (Figure 2.1), such as painting the models green,

Figure 2.1 Model Blitz 48 V

Figure 2.2 Main suppliers of TMHMI

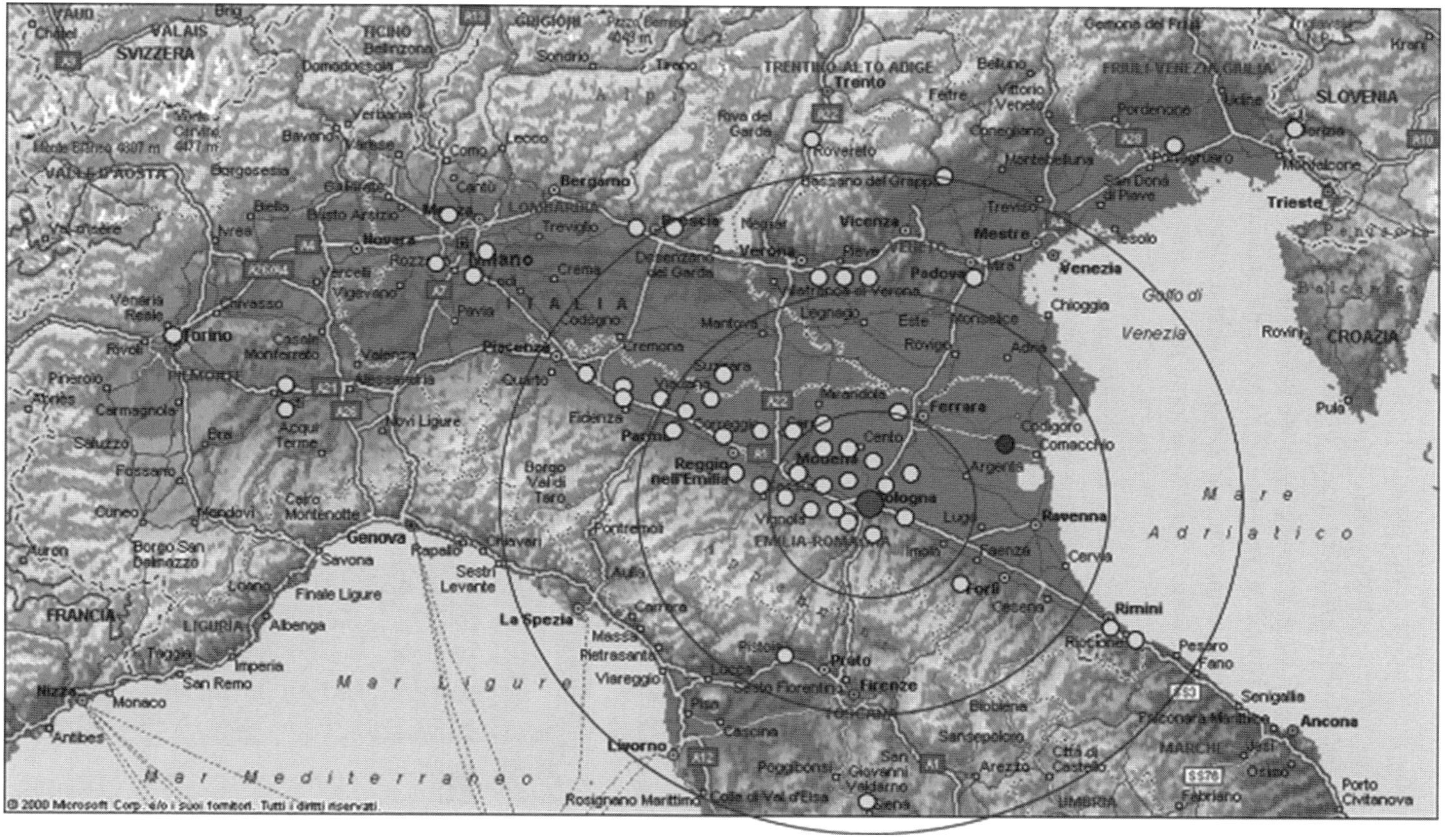

to remind customers of the company's attention to the environment and to make the product design stand out.

The owners subsequently decided to sell the company to BTP AB, a Swedish multinational company, market leader in warehouse material handling machines; a few months after acquiring CESAB, BTP AB was itself bought by Toyota Automatic Loom Machine. Basically, a chain of events that unravelled during the year 2000 brought a small, family-run company that sold 4,000 units a year to become part of one of the world's biggest conglomerates.

Another important factor for CESAB's early success had to do with the territory (Figure 2.2): Bologna is located in a highly industrialized area of northern Italy, with a strong reputation for mechanical engineering excellence.

Thanks to the strategic location, it is possible to create a qualified chain of suppliers within a limited area, capable of satisfying the most disparate requirements for mechanical and electronic parts. The company has maximized the opportunities offered by this network, gradually outsourcing all the non-core activities and focusing mainly on R&D, reducing the involvement in manufacturing to the point that in early 2000 it only had two mechanized stations, painting and fixed-position assembly lines for low volumes (Figure 2.3).

Figure 2.3 CESAB: Layout of the production area in 2007

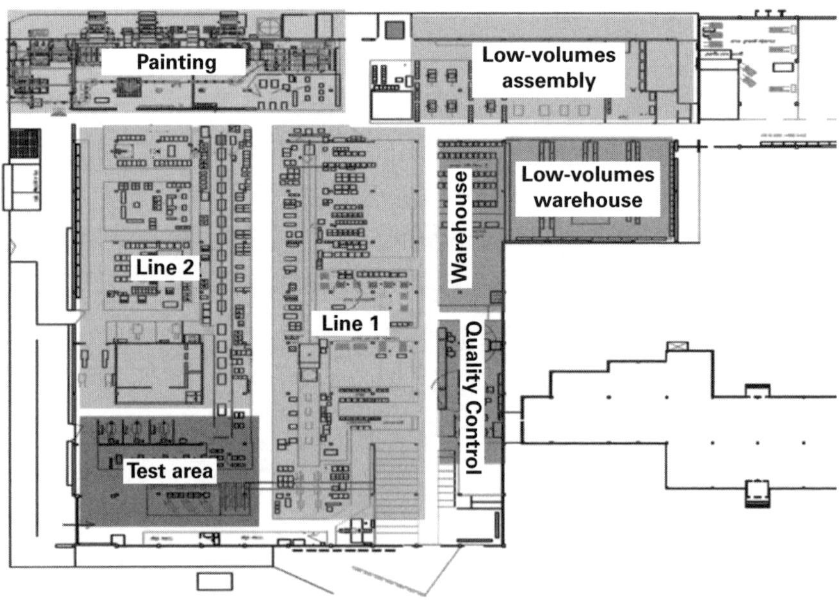

Figure 2.4 The entrance of Toyota Material Handling Manufacturing Italy

The integration with Toyota required the separation of CESAB's commercial brand from the Bologna factory; the site was therefore renamed Toyota Material Handling Manufacturing Italy in April 2016 (TMHMI, Figure 2.4).

03 The product

The counterbalanced forklift is an industrial product used for the handling of goods. Models can be classified by general characteristics, such as the type of power used (electric, diesel or LPG) and the loading capacity, usually indicated in kilograms or quintals. Customers can then choose among a wide range of options: different types of cab, varying mast heights to reach specific handling areas, assorted kinds of worklight, and much more, so that the trucks can fully satisfy their needs (Figure 3.1). Such a wide range of settings requires the use of a product configurator which, following precise rules, supports the customer during the creation of the most suitable machine.

The software of the product configurator processes the options selected by the customer, selecting and planning all the parts needed for the production of that model in the required configuration. The range of options is so wide that there is a very low recurrence of the same configuration; TMHMI analysed 6,000 orders of a single product family, and discovered that only 3 per cent of them had an identical setup.

The size of the product range that we offer to cover for any customer requirements, coupled with the extensive list of potential configurations, require a highly complex system of logistics and manufacturing (Figure 3.2).

In addition to the production of 'standard' forklifts, those that are chosen by the customers by selecting the options from a list, TMHMI also designs and manufactures 'special' forklifts. Those account for approximately 20 per cent of the site's total production, and are developed by R&D following

Figure 3.1 Current range of electric forklifts

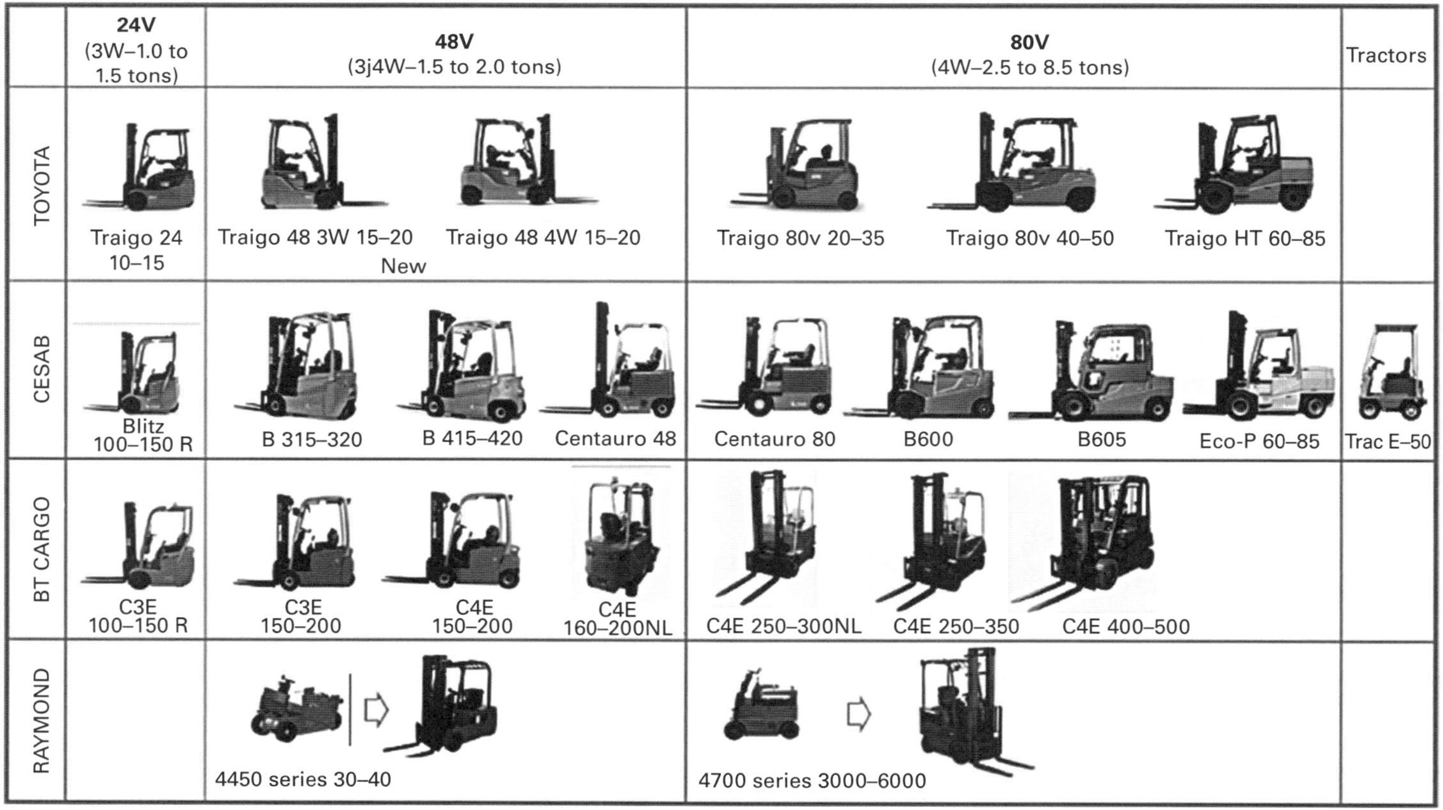

Figure 3.2 Internal combustion forklifts currently in production

	DIESEL/LPG (from 1.5 to 2.5 ton)	DIESEL/LPG (from 2.5 to 3.5 ton)	DIESEL (4.0 to 4.9 tons)
CESAB	 Drago 150–200	 Drago 250–350 E/H	 Drago 400–500
BT CARGO	 C4D/G 150–200 D	 C4D/G 250–350 E/D	 C4D 400–500 E

specific indications by the customers. When the existing configurations are not able to meet unique requirements, a department within the company is dedicated to the development of customized solutions. These products are added to the regular units and sent to the assembly lines, consequently extending the range of models that are manufactured at any given time.

04
From Push to Pull

As previously mentioned, the complexity of the product range and its countless configurations make any market forecasts extremely difficult, despite the considerable tools and resources made available by a global enterprise such as Toyota Material Handling. As a matter of fact, the reliability of forecasts is inversely proportional with the increase of customizable options. Moreover, we must also take into account all the orders placed by major customers, those who have purchased a considerable amount of trucks with similar or identical configurations, thus affecting the historic data upon which the demand forecasts are based.

This can only confirm and validate the assumption that the production and storage of machines, manufactured on the basis of forecasts that rely on historical data, involve the tangible risk of unsold units.

We previously tried to implement a possible solution as a compromise: a basic model was manufactured based on forecast and then put into storage. When the customer sent the order for a machine whose basic characteristics were matching those of the units in the warehouse, the forklift was retrieved and placed on the production line, where the assembly was performed according to the customer's specifications. This production model soon showed some limitations: manufacturing a forklift requires equipping it from the start with parts and features that are related to its final configuration. For instance, chassis and counterweight depend on the load capacity, while the wiring is designed to be adaptable to a limited set of configurations. The same issue applies to the hydraulic system, whose number of motions relies on the mast and on possible additional devices that the customer has included in the configuration. This method, developed as a compromise between a manufacturing system based on mass production[1] (where the setup needed to switch between productions required downtime and massive amounts of labour) and a system that could rely on the choices made by the customer, has shown its limitations. Not only did we have an issue of compatibility between parts and options, but the final configuration of the forklifts took much longer than a traditional all-in-one assembly.

Furthermore, continuously moving the trucks between the warehouse and the assembly line was a constant time-wasting exercise. Finally, the increase of the range and of the available options, indispensable for competing in our reference market, affected the reliability of forecasts and reintroduced the potential risk of getting stranded with unsold units.

In order to avoid freezing a vast amount of capital in the warehouse, the most effective solution was to manufacture the machines following customer orders, switching from *Make To Stock*[2] to *Make To Order*.[3]

Such a transformation required a deep corporate shift, not only concerning the planning system, but also affecting the organization of assembly lines, logistics and supply chain.[4]

This strategic plan started at the beginning of the new century, right after the company became part of the Toyota group. At the time, the limited number of models and sales could allow for an organization that manufactured by single order; although it would have been inadequate for today's range of models and options, it represented the first, necessary step to kickstart a TPS-oriented transformation. Today, Bologna's production site manufactures 24 different models; six of those are manufactured on a single line and represent 90 per cent of its yearly output, while the remaining models are produced on a 'Big Size' dedicated line.

The first step was to switch to Make To Order: it was the most important decision for the transformation of the site and the implementation of the Pull[5] approach, in line with one of the basic principles of TPS: make only what the customer wants.

With a Make To Order approach, the completed units are stored in the warehouse only for the time it takes to organize the delivery to the customer, therefore drastically reducing the need for storage space, improving the company's cash flow and virtually removing the risk of unsold units.

When in 2008 the Lehman shock cut in half the European market for material handling, causing sales to crash from 440,000 to 200,000 units, the system flexibility proved to be invaluable. In order to adapt the output to a dwindling demand, the company constantly adjusted its production quantity with the support of the supply chain; this avoided the deadweight risk of forecasts, which always take longer to align with real market conditions.

If we had decided to manufacture the volumes based on forecasts, without adjusting our capacity to actual demand, we would have manufactured a considerable amount of trucks that would have had to be stored for a long time because of a lack of buyers, badly affecting the company's P&L.

Notes

1 As opposed to Lean production, mass production implies the manufacturing of a single standardized product series in large amounts. The classic example is Henry Ford's assembly chain and his famous statement regarding the company's Model T: 'You can have any colour you want, as long as it's black.'

2 Make To Stock (MTS) is the manufacturing of standard, simple products on the basis of sales forecasts, destined to be kept in storage. They are usually goods with limited value per unit, for which there are significant market outlets.

3 Make To Order (MTO) is the system where production is carried out based on the customer order, while planning and engineering can be done in advance. The choice of which production method to implement depends on the company's positioning in the market; the decision relies on analysis conducted by operations, marketing and several other company departments.

4 Management of the supply chain, including logistics.

5 The Pull approach states that the processes are 'pulled by the customer'. It is opposed to the Push approach, where processes that include production are based on forecast and/or batches.

05
Production lines

The implementation of Make To Order has surely contributed to eliminate the risk of surplus units; however, it required completely rethinking the production system, starting from the assembly lines. The orders reach the site daily, and at times are considerably different from the averages of past production, both in terms of volumes and models. It is not uncommon for weekly orders to be more than 50 per cent different from what was forecast.

For this reason, maximizing the use of assembly lines is a delicate and complex process, which forced the company to take two crucial decisions:

1. reduce the number of assembly lines;
2. reduce setup times to implement a one-piece flow production model.

Setting up the assembly lines for macro products is a widely adopted layout[1] for manufacturing companies, because it allows:

- good optimization of the lines in terms of assembly efficiency, meaning high saturation of the time cycles allotted to the stations;
- simplified logistics, thanks to a reduced quantity of codes to manage along the production lines;
- simplified assembly procedure, which allows for simpler training for the operators;
- assembly quality.

On the other hand, this system complicates the activity of volume allocation to the production lines. Let's assume that we allotted:

- a daily output of 10 units to the line of product A;
- a daily output of 15 units to the line of product B.

If the orders coming from the sales department vary from the settings above, the capacity allotted to the lines must be adjusted to meet customer requirements. This implies frequent changes, which means investing resources to implement them. Another downside is the space needed to fit a high number of production lines, one for each model currently on offer, which also implies higher levels of investment and the likely duplication of some indirect functions.

In order to overcome those hurdles, TMHMI decided to manufacture the best part of its output, around 90 per cent of it, using just one of the assembly lines installed in the site. If we consider the example mentioned above, the daily production line capacity remains 25, regardless of the kind of model being assembled.

I often wondered whether saturating the capacity of a single production line was the way to go, while the other line, which was identical, spent most of the time on standby. Time has told us that it was the right choice, as the efficiency level can testify: a line that made 25 forklifts a day now has an output of 60, with the outlook of reaching 70 in the short term. It is important to notice that this result was achieved without substantial capital investments, optimizing the workstations on a daily basis following the *Kaizen* philosophy. With this system the technical capacity of the lines, or the maximum theoretical capacity, has improved constantly, regardless of the real demand coming from the market.

As far as production and process engineering are concerned, our target was to optimize processes and improve the layout and efficiency of the assembly stations. Excess technical capacity was highlighted by leaving workstations unattended along the line. This was another key factor for TMHMI's growth: for a global business, excess capacity implies the ability to introduce new models while keeping investments to a minimum.

I still remember when our Japanese parent company had to choose the European site that would manufacture the new 48V model, designed for the most important segment of the European market. Top management from Toyota Material Handling visited the factory several times, to assess the suitability of the plant for a substantial increase of the production volume, and to estimate the investments that would be needed to achieve such a target. Showing them empty stations was one of the most strategic and successful choices that we made, because it allowed us to prove that the line had the potential to perform as required. We had to sell it to them; after all, a global company has to deal with two kinds of competitors: those outside and those within the group.

The strategy that we adopted bore fruit: today TMHMI uses line 1 to assemble large-volume models with smaller labour requirements, saturating it at above 80 per cent of its total capacity. Line 2 has been dedicated to the assembly of forklifts of bigger size, which sell in smaller volumes and need more hours of work (Figure 5.1).

Figure 5.1 Layout and models produced

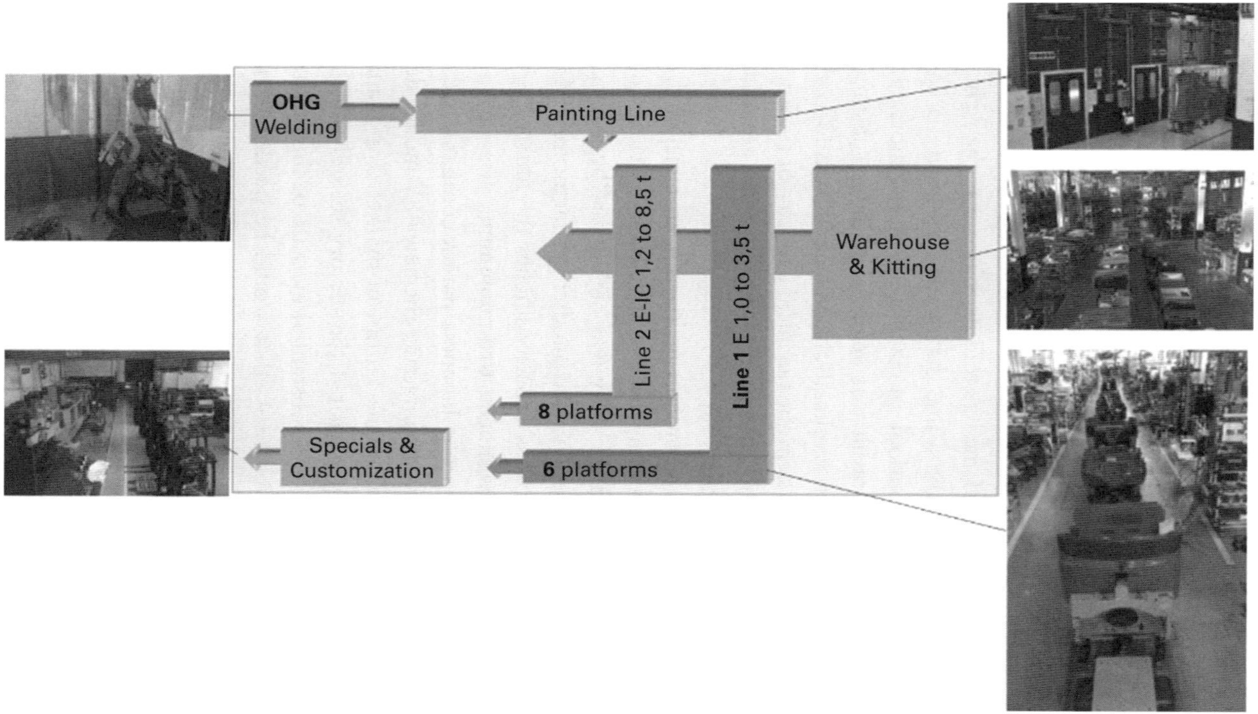

It is fairly understandable if people assume that batch production is a more efficient system. However, it only works in certain environments. Let's consider a task such as sending 100 letters: most of us would start folding all the sheets, then writing all the addresses, placing all the stamps, stuffing the letters in the envelopes and closing them. If we think of doing each of those actions one by one for every single letter, we rightly feel like we are being inefficient.

However, this method is ideal only if all the letters are exactly the same, and we already have 100 recipients; if that is not the case, for example if the recipients are just 50, or if their address has changed, you risk filling your warehouse with useless letters, or disposing of some of them because the addresses are wrong. Saying that mass production is never efficient is incorrect, but there are very few markets that suggest the adoption of such a production setup.

Let's make another example, this time with crayons. The analysis carried out with production and management control suggested that the ideal production batch is 10 units per colour. A customer then orders one red crayon; we produce a batch of 10, 9 of which go to the warehouse, and we deliver one to the customer. In the meantime, another customer has ordered a green crayon; we delay production until the batch of red crayons is ready, then we switch to the production of 10 green crayons, 9 of which will again end up in storage. This simple example highlights how, in a market that is not willing to wait and expects a wide range of choices, a system based on this approach is clearly unsuitable to satisfy the customer; the ideal system would be to manufacture crayons of several colours with a production batch of 1 (Figure 5.2).

Transferring the production of most of the models on a single line is still not enough, unless it is possible to overcome the constraints of minimum batch production for each model, organizing the line so that customer orders can be fulfilled by a single-unit batch procedure.

Figure 5.2 Batch manufacturing versus one-piece flow

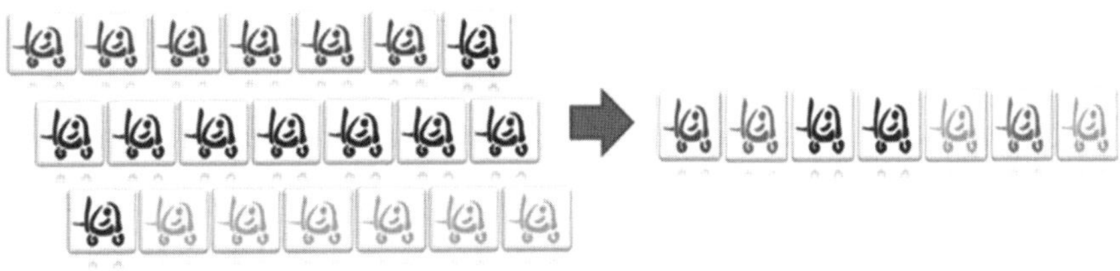

The strategic choice here is to evolve production and logistics in order to achieve maximum flexibility: manufacturing has to adapt to customer requirements, not the other way around. In order to organize production lines so that they are suitable for the assembly of such a wide range of products, with substantial differences in assembly times, it is critical to carry out important modifications in the following areas:

a times and methods;

b *Heijunka*;

c production engineering;

d organization;

e factory logistics.

Let's begin with times and methods along the assembly stations. One of the main hurdles to overcome while applying the one-piece flow system is the need to balance production lines. As far as TMHMI is concerned, various models and customer options can imply substantial variations of production times. In order to maintain a steady workflow for the operators along the line and manufacture all the various models in sequence, we created sub-groups where some of the parts of the models that require more time are pre-assembled (Figure 5.3). Those sub-groups are located next to the line: once the model is pre-assembled, the item travels by roller straight to the station that will take care of the actual assembly.

Further support to the one-piece flow system comes from the *Heijunka* philosophy (*Hei*, 'without anomalies', and *Junka*, 'levelling'), aimed at levelling the production of the plant. The goal is to balance workloads throughout the stations, alternating models with different labour requirements. The sequence is prepared by manufacturing planning, which drafts an assembly schedule that complies with the requirements (and takes limitations into consideration) reported by production. Such a working method also applies to the sub-groups, so that they are aligned with the maximum capacity of the area. For example, the sub-group area dedicated to the assembly of the cabs has a daily capacity that depends on the type of cabs chosen by the customers. Full cabins with heating and air conditioning require a lot more labour than cabs equipped with only front and back windows. Assuming that the amount of people working on the area remains the same, the maximum daily capacity taken into consideration at the planning stage depends on the quantity of forklifts with cabs, and their characteristics. Production engineering has reviewed,

Figure 5.3 Pre-assembly area

assessed and developed all the equipment required for manufacturing, so that it can be easily adapted to the entire product range; an example of this approach are the screw guns, whose specifications make them suitable for a wide range of screws and torques. Other elements, such as the hooks for handling components, have been developed in-house to allow for maximum flexibility.

Training of operators and the *training dojo*

Our organization focused on training the operators to manufacture different models. For this purpose, a very important function was performed by the areas called *training dojo* (Figure 5.4); *dojo* means place (*jo*) where you follow the way (*do*). These are proper training areas, specifically set up so that the operators can practise and learn the use of the main equipment they will use on the line, assembling critical components, simulating quality issues and improving manual skills.

Figure 5.4 The *training dojo*

Training dojos are used to train:

- new hires;
- operators who are switching tasks;
- operators whose performance needs to be improved.

For instance, in order to train operators on their manual skills, dedicated boards have been installed for practising their speed and precision, following a specific route with a rope (Figure 5.5) or tightening screws in different locations (Figure 5.6).

The *training dojo* is also implemented for training operators on specific, quality-related elements, in particular those that involve the assembly lines. One of the most complex issues that we had to solve, for example, is how to deal with wiring connections; these are performed with connectors that are pushed one inside the other to ensure the correct flow of electricity. Although the connection looks properly done and the connectors are properly placed, at times the internal cable does not link properly. A dedicated board has been created to solve the issue (Figure 5.7), where operators are trained to spot immediately this kind of fault during assembly.

Figure 5.5 Manual skills simulator

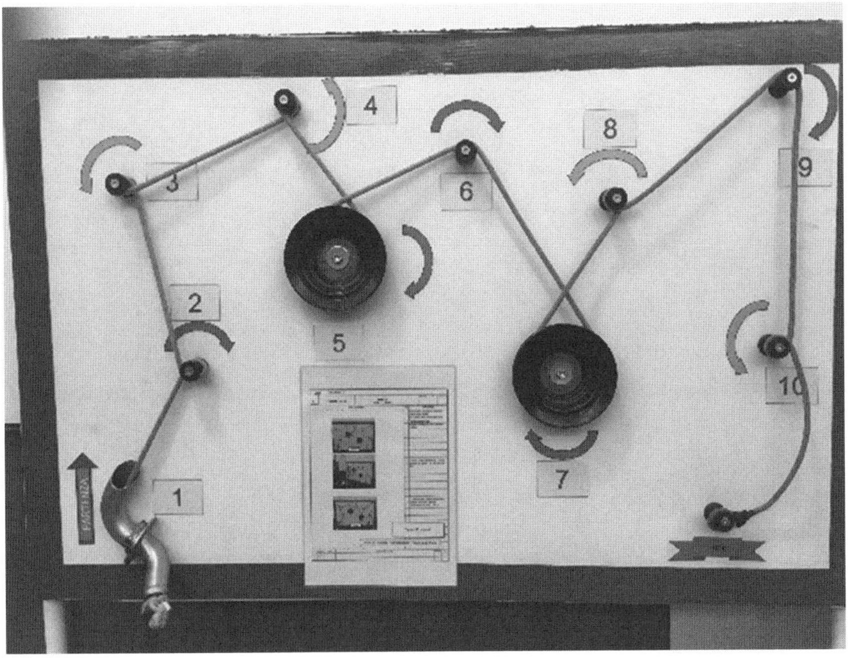

Figure 5.6 Screw tightening simulator

Figure 5.7 Wiring connections training board

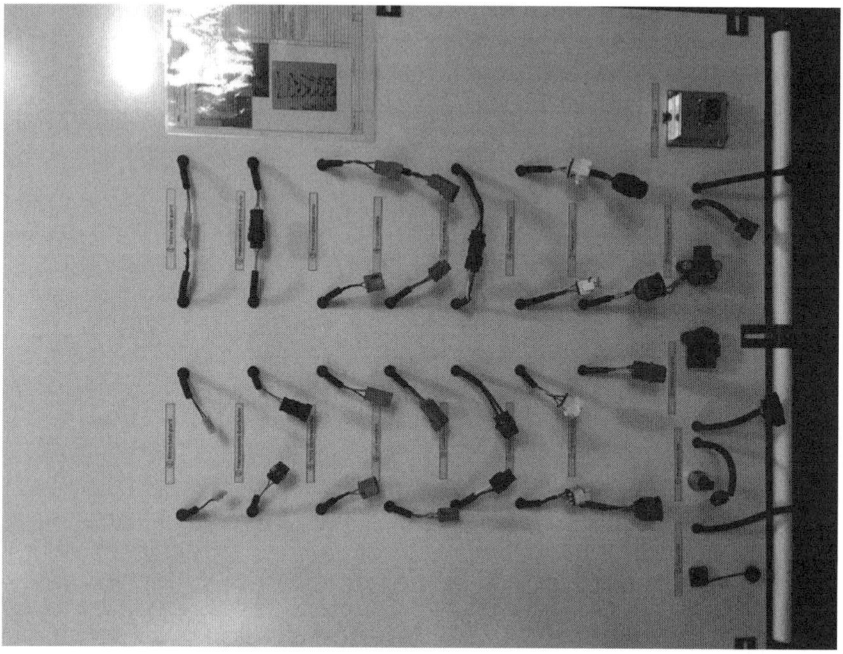

Operators' training follows a specific procedure which, for new hires, begins on their very first day, with a general introduction to the company, a 'welcome on board' booklet, where they can find all useful information (from the practical details of the organization to safety instructions, including the history of the company). The booklet immediately introduces the main concepts of TPS. Training ranges from a minimum of three weeks for the simplest workstations, to four to five weeks for the most complex duties.

- **The first week is based on theory**, where the future operators learn a basic set of notions regarding assembly procedures, safety and TPS, aided by practical sessions at the *training dojos*. For this purpose, in addition to the *dojo* created for work procedures, there is also one that focuses on safety and environment, illustrating all the safety-related risks with examples and practical tests. At the end of this stage, the new hires are tested to assess their knowledge level.
- **Only the second and third weeks focus on training 'on the job'**. The new operator, supported by a trainer, begins to learn directly at the workstation where they will be assigned.
- **After the third week**, if the operator has received a positive judgement, **they can start working at the station.**

The training steps described above are displayed on a board placed by the reference area. A dedicated graph, which splits the skills into four sections, was implemented to identify the progress stage of the employee's training (Figure 5.8).

Finally, the plant's logistics have been completely overhauled; the topic is discussed separately in Chapter 8 due to the complexity and relevance of the operation.

The 'one-piece flow' principle was also implemented for processes that are less flexible by nature, such as those carried out by the welding and painting lines. In this regard, we opened the carpentry department in 2010 to develop a flexible system for welding the driver protections (Figure 5.9). The project had the goal of creating a new department that could comply with one-piece flow and weld over 50 different types of driver protection units for our range.

Figure 5.8 Training and skill stages of the operators

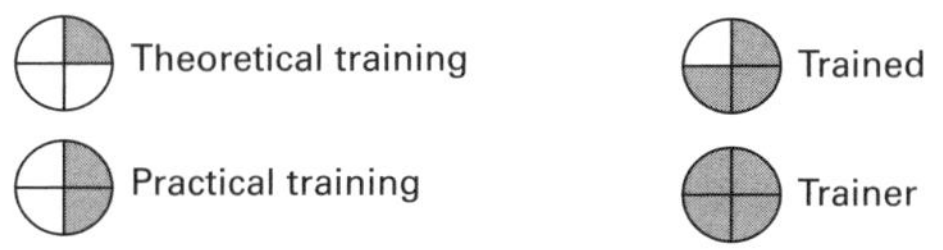

Figure 5.9 Rendering of a driver protection unit

One of the project imperatives was that any potential solution had to allow for the shortest setup times[2] when switching models; moreover, the station had to be:

- tightly set up, to minimize the movement of people and materials;
- flexible and simple, so that it could be adapted to support future models.

It is worth focusing on the idea of 'flexible and simple': these are characteristics that featured in all my visits to our parent factories. A visitor who enters a Toyota factory in Japan, expecting to find a plant where technology and complexity predominate, is bound to be disappointed. Toyota is nothing like that, not even in the automotive sector, where volumes and efficiency play a fundamental role.

In one of my visits to Japan I had the opportunity to see the site where the Land Cruiser is made. This production site, considered by the group as one of the best for cleanliness (5S) and efficiency, manufactures a car every 2.5 minutes. The manager of the plant showed us many of the improvements they implemented. One of them, which was suggested and carried out by the workers themselves, involved the creation of a system that could gather from a container the exact amount of screws needed for the assembly at a specific workstation. In that case, only three screws were needed for the operation, but the line speed and the use of gloves made the screws very hard to pick. The small team of operators solved the issue by attaching three magnets at the end of three ropes (Figure 5.10).

The ropes are all tied to one end of a balance beam, while on the other end sits the screw gun used for the operation. When the operator picks the screw gun, the beam tilts and the magnets reach inside the box of screws

Figure 5.10 Example of innovative equipment by the assembly line

placed right below them. As soon as the screw gun is placed back on its seat, the beam lifts the ropes, carrying the required screws, which can be easily picked by the operator. I can only imagine what could have happened if the same issue was faced by a European company: they would have studied a complex solution that required considerable investments without matching the same level of reliability.

It is also important to remember that workers are keener to accept a new solution if it is created by themselves or by their peers; it contributes to their motivation and makes them feel involved in the company's process of continuous improvement.

Inspired by the principle of reaching the ideal balance between technologies and simple solutions, focusing on *Kaizen* ideas suggested by the individuals, we have developed a new welding department dedicated to driver protection units. To ensure that the plant was simple and functional, the area was divided into three parts: tacking, welding and finishing (Figure 5.11).

Tacking[3] is the operation where the parts of the driver protection are connected with tack welding to form the shape of the unit. It is by far the most complex process, since it must ensure the quality of the protection shape, guarantee the correct size, allow flexibility so that the unit can be adapted on multiple models, and of course it must guarantee complete reliability. We studied the techniques implemented by other businesses that used similar products and processes, and we noticed that one of the most common solutions was the use of fixtures, where the parts are assembled and welded by robots, eliminating the tacking stage altogether. A more detailed analysis of

Figure 5.11 Carpentry department layout

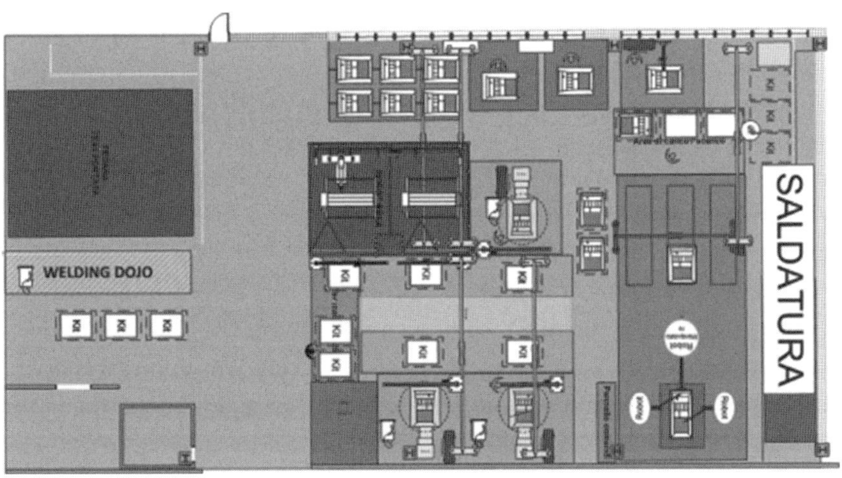

this system highlighted some issues, particularly regarding flexibility and complexity of the equipment. The development of a fixture that could ensure quick setup times, together with the need to allow adequate room for the robot arm, not only required substantial investments from our part, but it also needed us to make some compromises in our philosophy.

We found a way to adapt different approaches and create our own: we adopted fixtures with manual holding and positioning systems, and weld tacking is performed by an operator, who can reach all the critical areas much more easily than a robot. This method relies on the flexibility of the human body, avoiding complex software and the equipment that would be needed by the robot arm, since the operator can easily reach the most remote areas that are more difficult to weld.

Welding, which must ensure speed and constant quality through repeated motions, is performed by two robots (Figure 5.12). In order to ensure that the welding torch is always perfectly positioned and that it can reach all the areas, the unit is moved by a manipulator: since the structure is already tack welded, handling is easier and guarantees total freedom for the robot motions. The manipulator retrieves the protection units and places them at the end of each on a buffer area, where three units can be temporarily stored, so that the robot can operate continuously.

The finishing stage, where the welding is checked and any imperfections are addressed, is manually performed by the operators.

The flexibility offered by technological solutions must also be supported by logistics, which must supply the parts required for the assembly of the protection unit. We found that the creation of a supermarket area was the most suitable solution to achieve this goal. The area stocks all the parts for the assembly of driver protection units; an operator prepares each kit according to the daily production schedule, and the kits are then placed next to the tacking area. The parts are restocked from the external supply chain with a Kanban system.

Finally, in order to minimize the movements of the driver protection units when moved around the plant, all handling is carried out with the same trolley, internally developed by process engineering; the trolleys are also pulled by a towing tractor[4] and used to transport three units at a time to the painting area (Figure 5.13).

These simple solutions suggested by the operators helped to develop the **new carpentry department**, which had an initial target of 38 driver protection units per shift, with the capability of producing two different models. Now, thanks to the continuous *Kaizen* improvements and without significant

Figure 5.12 Carpentry department, robot cell

Figure 5.13 Driver protection unit transport system

investments, the line assembles 58 units per shift, equipping three different forklift models.

The same one-piece flow approach was brought into the **painting facility**, where the robots were programmed and optimized so that they could work on any part, regardless of colour or shape. Our product range uses four different colours, two shades of grey, a green and an orange; this made the application of one-piece flow very complicated: switching colours requires waste of paint and the use of solvents to clean the equipment, which means higher costs and environmental impact. While these limitations suggest that proceeding with single-colour batch painting is the most suitable approach for this department, we found a compromise by balancing the number of colour changes with the number of batch-painted chassis and protections. Now the system implies four hours of production of batches with the same colour, but studies are being conducted to reduce the batch to two hours. Sometimes TPS appears to be in contrast with common sense; however, an *ex post* analysis that assesses the efficiency of the whole system, rather than of a single unit, shows that this approach is unquestionably more efficient.

Gradually reducing the batches or the intermediate production buffers was a very effective method to introduce the cultural shift in small stages. Years of mass production logic convinced everyone that the best way to avoid disruptions for customers was to keep high quantities of products in stock. Every time we recorded a shortage, we thought that the easiest solution was to increase production. When we started to gradually reduce our stock, our department managers suffered from mild anxiety; however, a gradual reduction gave time for people to get used to the new situation, while they could see for

Figure 5.14 The seven kinds of waste according to TPS

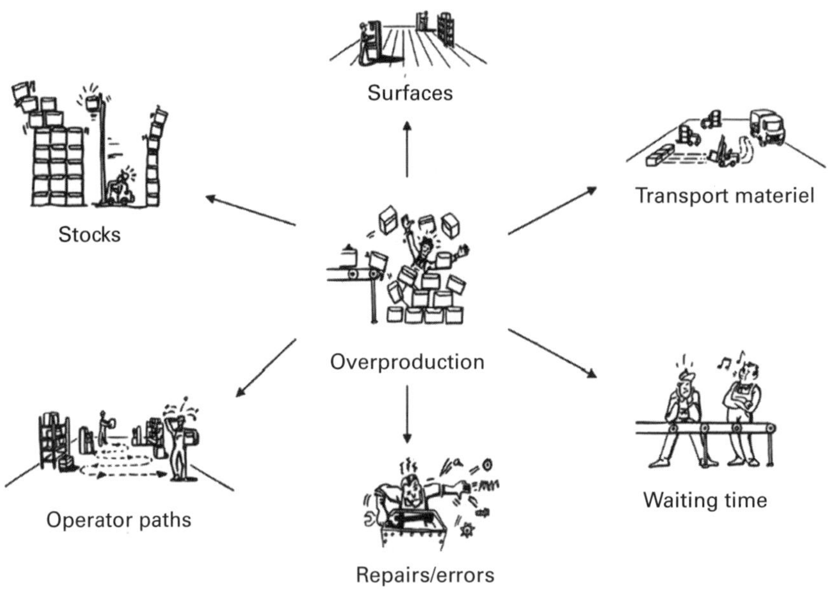

Figure 5.15 TMHMI On Time Delivery graph

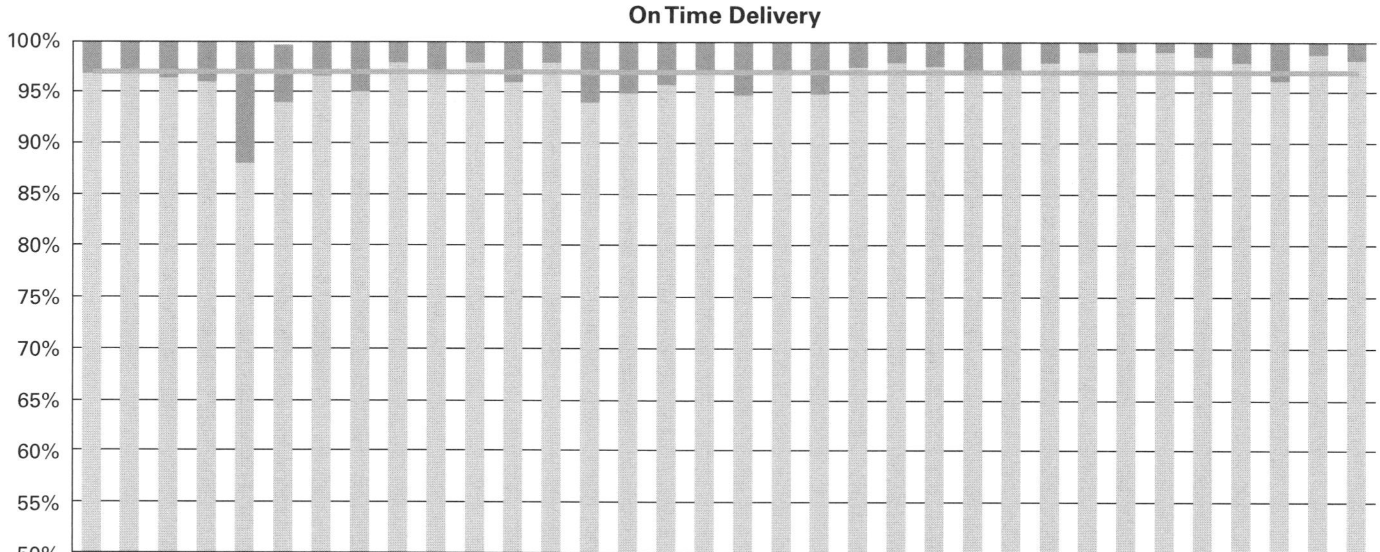

2014	1	2	3	4	5	6	7	8	9	10	11	12
Late	3%	3%	4%	4%	12%	6%	3%	5%	2%	3%	2%	4%
On time	97%	97%	96%	96%	88%	94%	97%	95%	98%	97%	98%	96%
Target	97%	97%	97%	97%	97%	97%	97%	97%	97%	97%	97%	97%

2015	1	2	3	4	5	6	7	8	9	10	11	12
Late	2%	6%	5%	4%	3%	5%	3%	5%	3%	2%	2%	3%
On time	98%	94%	95%	96%	97%	95%	97%	95%	98%	98%	98%	97%
Target	97%	97%	97%	97%	97%	97%	97%	97%	97%	97%	97%	97%

2016	1	2	3	4	5	6	7	8	9	10
Late	3%	2%	2%	1%	1%	1%	2%	4%	1%	2%
On time	97%	98%	99%	99%	99%	99%	98%	96%	99%	98%
Target	97%	97%	97%	97%	97%	97%	97%	97%	97%	97%

themselves how the level of service was actually improving in the process. TPS has, not by coincidence, classified seven different cases of *Muda*[5] (waste), and identified overproduction as the cause of all other waste (Figure 5.14).

Over time, the cultural shift brought by TPS has changed people's perception: the stock, once a safety element, has become a *Muda*. The whole site has joined forces to reduce the stockpiles of semi-manufactured and finished trucks it was storing; internal lead time, or the average time from the beginning of the production phase to delivery of the finished product to the warehouse, and the WIP (Work In Progress) have become two crucial key performance indicators (KPIs). Today, with a daily output of 75 units, the WIP (the total of all the forklifts inside the plant) is 140, which means less than two days' production.

Notwithstanding the constant effort towards the reduction of WIP, the level of performance and service offered to TMHMI customers has improved constantly, so much so that production hits 98 per cent of the forecasted delivery dates for all manufactured trucks (On Time Delivery).[6]

Notes

1 Organization and setting of an area, including the location of equipment.

2 Setup is the time required to perform all the operations during downtime to switch production from model A to model B (ie mould and equipment change, removal and replacement of the parts no longer needed, etc).

3 During the tacking stage, little tack welds connect all the different parts. The process is usually supported by masks that ensure the correct shape of the unit; once finished, the unit can be moved without altering its structure, ready for the final welding stage.

4 Towing tractors are small towing vehicles that allow the movement of the trolleys inside the plant. They are widely used in TPS systems, because they are very efficient at transporting a considerable volume of parts.

5 *Muda*, the Japanese term for waste, is one of the three kinds of waste that are contemplated by the TPS philosophy.

6 On Time Delivery measures the amount of delivered forklifts within the first delivery date given to the customer. When Toyota Material Handling production sites receive a customer order, they issue an order confirmation indicating the delivery date for the order.

06 The planning template

Before we focus on some of the unique elements that are part of TMHMI's planning template, it is useful to give a broad introduction of the system. As previously mentioned, the orders are received daily from the sales network of TMHE (Toyota Material Handling Europe), mainly composed of commercial branches owned by the group, which covers the entire European market. The sales network has over 40 units between companies and importers, sending their orders directly to the production sites on a daily basis. Each country has its own particular market, with different volumes and best-selling models; such diversification means that the daily orders can vary from a few units to hundreds. In order to keep the plant's capacity as stable as possible without having to resort to frequent takt time[1] changes, we maintain an order backlog[2] that allows the lines to be supplied with a fairly constant daily volume. With this method, the backlog fluctuates according to the orders received; since the daily output does not vary, the delivery times of a forklift depend on the actual orders received.

The time that elapses from the reception of the order to its delivery is called Lead Time (LT). Since LT is a sensitive item for sales, it must remain within a certain range, varying between 6 and 10 weeks depending on the model to be manufactured. If LT falls within this range, we do not interfere with production capacity, which is raised or lowered only when LT hits minimum or maximum levels. Yearly and monthly forecasts are used to assess whether the predefined production capacity is in line with LT targets and BDG volumes received from TMHE head office. Forecast, LT and production capacity are added in a database whose graph (Figure 6.1) displays the trend of the inserted values.

The width of options and models does not allow for the adoption of a 'frozen' base for each finished product. Based on the model and options chosen by the customer, a product configurator processes the requirements for each

Figure 6.1 LT trend related to production capacity and orders

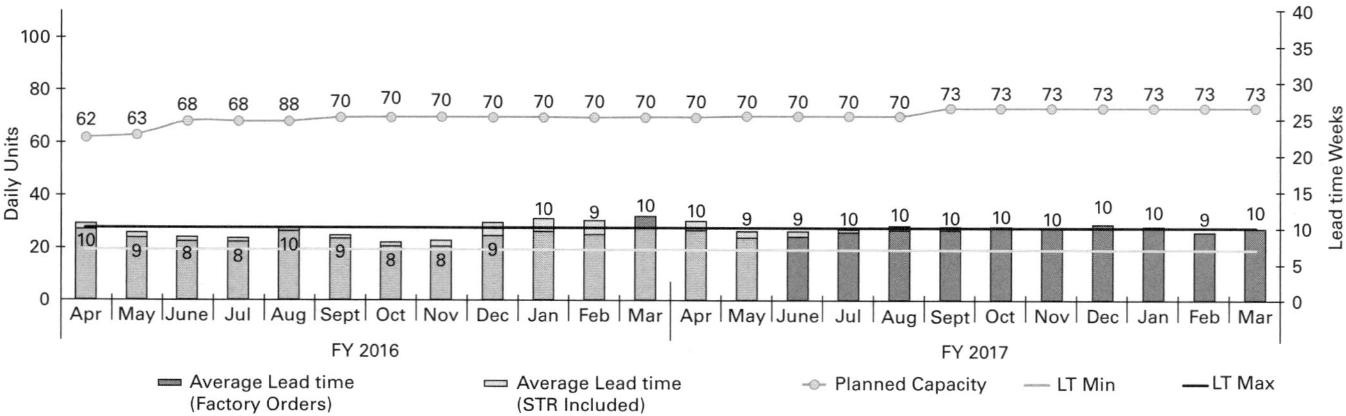

specific item, creating a job order. The MRP[3] finalizes the delivery plan each week, which is then sent to the supply chain. The following two weeks of production are 'frozen' (Figure 6.2): no changes of quantities or component specifications are allowed. From the third week it is possible to apply modifications, which increase in number with the passing of time.

Using one-piece flow allowed for complete planning flexibility, since there are no limitations to the potential numbers of models and options. The only restriction lays on the daily production capacity of the lines, which is saturated with the daily list of orders. With this system, any modification of the production mix set by forecasts does not create any particular problem to the saturation capacity of the lines, while at the same time ensuring that the customer is given the optimal Lead Time. High flexibility of the two production lines also gave us the opportunity to develop different planning systems for the materials, reducing the stock of raw materials to a minimum and optimizing production capacity.

The main planning system for TMHMI is always the MRP, which creates a delivery plan based on the backlog of current orders and on the forecasts for the incoming months, covering a timespan of about one year. The supply chain receives a weekly delivery plan which includes a frozen and a forecasted period. Some of the typical parameters of MRP have been reviewed and amended, so that they can be adapted to the new requirements of flexible production systems. For example, if we take into consideration the minimum batch[4] factor, a normal one-piece flow might define the purchase of

Figure 6.2 MRP applied to TMHMI

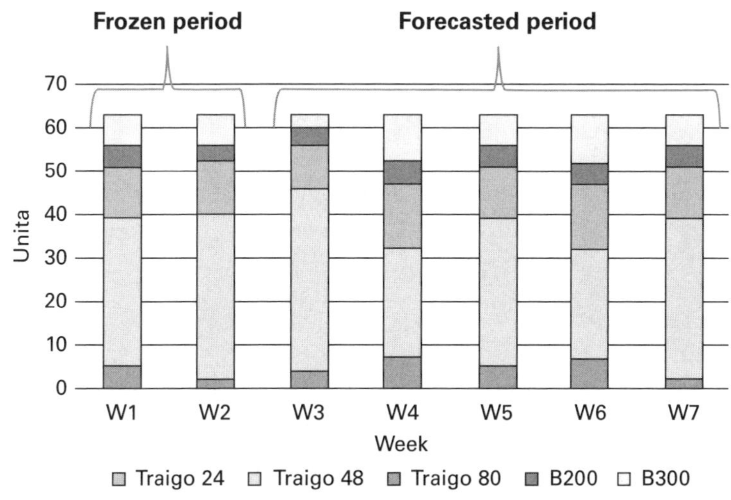

materials following the minimum batch requirements for every single order; this would mean the need to store a considerable volume of materials.

MRP, which is a system that is normally considered as a Push model, has been integrated with the typical methods of TPS, a Pull model by nature. The first action that we implemented to balance the stock of raw materials and the actual usage by the production lines was the implementation of **Kanban**. *Kan* (看) means 'visual', while *ban* (板) means 'signal': it is the most widely known planning method for materials. There are several techniques and rules for its use. In TMHMI, each container with the same item contains a Kanban card (Figure 6.3).

When the operator retrieves the last part from the container, they pick the Kanban card and place it in a collection box before moving to the next container. Logistics periodically gather all the cards and proceed with the purchase of the materials. Restocking can be performed using the internal warehouse (internal Kanban) or the suppliers (external Kanban). MRP was integrated with the Kanban system: MRP gives information and sets future purchases from the supply chain for the medium to long term, while Kanban deals with the manufacturing and supply of the materials indicated on the card.

The supply chain, considering its processes and production strategies, can identify its own decoupling points, which means understanding when to keep on manufacturing based on the planning received by MRP (a Push system), and when to activate the typical Pull stage of Kanban. For instance, a company which follows the logics of mass production will define its production volumes following MRP requirements; the finished goods

Figure 6.3 A Kanban card

warehouse shall be the last stage from which to ship the materials required by the orders received via Kanban.

Toyota's philosophy regarding the supply chain involves an active support of the suppliers, so that they can develop and implement their own TPS methods. This derives from the fact that the supply chain is seen as a long-term business asset; any element that might increase efficiency and offer a competitive edge can strengthen the relationship and the performance of both businesses. It is much preferable to making large investments with a new supplier so that it can meet the group's quality and method targets, without considering the risks related to quality.

Another planning method that was developed by TMHMI to optimize purchases is the reorder point. This system is usually implemented with materials that are low cost by nature, such as adhesives. Although most of them are extremely cheap to purchase, their shortage can at times disrupt operations. When the operator retrieves the first unit of adhesive from its container, the item is classified as understocked and must be reordered. The same principle has been applied to all small hardware parts.

As previously noted, the introduction of Kanban has ensured great results with stock rotation, balancing the quantity of materials stored inside the plant at any given time and simplifying management operations, which are now shared by planning and logistics.

This considerable step forward, however, did not completely solve the issue: there was still a huge amount of materials stocked, especially heavy and cumbersome items, which caused several issues for logistics concerning handling and the need for large storage areas; moreover, it meant that a substantial amount of capital was kept frozen in the warehouse.

Logistics had to deal, for instance, with ensuring that the oldest materials were the first to be used, so that quality and obsolescence issues could be avoided. Several attempts were tried to implement a FIFO[5] system (First In, First Out), but the space and handling requirements of crucial components such as the frames made the operations very difficult. Moreover, the growth of volumes and models produced in the plant caused an increase of mistakes in the retrieval of the parts: logistics issues caused by common errors.

Our visits to the production units in Japan and the USA made us discover what became a game-changer for our planning system, called *Junjo*. I must admit that working for a large, very 'factory-centric' group has the advantage of offering a huge volume of information from the other sites, from which to find and implement tools and solutions that have previously been applied with positive results. The Japanese call this process *Yokoten*,[6] and it is widely adopted by the group. *Junjo* was already used in other plants, and we immediately understood that it could represent the ideal solution for our problems.

Junjo means 'sequence'; it consists of a planning system that brings considerable simplification and efficiency within and outside the plant. Planning allocates the daily production capacity based on each order received, until it reaches full saturation. Subsequently, MRP plans the purchase of materials and issues the delivery plan to the supply chain. Close to the start of production date, each forklift that will enter the line on the same day is given a sequential number, from 1 to the specific daily capacity. Each day is marked with a different colour. With this simple method, each truck planned for production has its own, unique reference number, making for easier identification (Figure 6.4).

This process is undertaken on a daily basis, planning the production sequence of the forklifts that need to be assembled; each day, the sequence for the following seven production days is frozen. The software processes the information and defines the materials required for the production of each number, notifying the supply chain. Suppliers will deliver only the parts needed for the required daily sequence, following the same sequence of the line itself. The same approach is applied by all the plant's internal processes, starting from carpentry all the way to the assembly line.

Thanks to this system, the plant receives only the parts needed for the operations planned for the day. The system contributed to reduce the volume of materials stored in the warehouse, an important advantage when it comes to items with high value; it has raised the turnover rate of the

Figure 6.4 An example of a *Junjo* card

Allotted day	XX/XX/XXXX
Sequence no.	4
Production order	87717C
Part code	51601T106071

Daily colour code

Monday	Tuesday	Wednesday	Thursday	Friday

components and has consequently reduced the area needed for storage. Any production issues that prevent the completion of the daily sequence are re-planned outside the seven-day period, after which all the stock levels managed by *Junjo* and the sequence numbers are realigned with the first forklift that will be assembled, again marked with the number 1.

Although this method drastically reduces the safe stock levels, the quality of service of the suppliers that have adopted the *Junjo* delivery plan has improved immensely, hitting almost 100 per cent in terms of delivery dates and quality of the parts supplied. I personally justify this success with the fact that now *Junjo* directly connects the supplier with the production activities of the plant, as if it were an external production unit. A simple and clear process makes the supplier aware of the consequences of any disruptions along the chain. The success of this planning system also meant freeing areas that were previously destined for storage, creating new spaces and opportunities.

On top of our planning systems, another important contribution was given by *Genchi Genbutsu* ('go to the source'), a system where the personnel in charge of materials planning is brought directly into the production areas. Seeing the quantity of parts with one's own eyes is much more efficient than any data processing. Often during these visits I would use my index finger to indicate how much dust had gathered on the parts that had spent a long time in storage: I picked up the dust, showed it to the visitors and told them that it was our real 'turnover index'.

Quoting the words of Taiichi Ohno, one of the founders of TPS, 'the higher the quantity of stock in the warehouse, the lower the chance of finding what you need'.

Figure 6.5 Turnover index of TMHMI warehouse

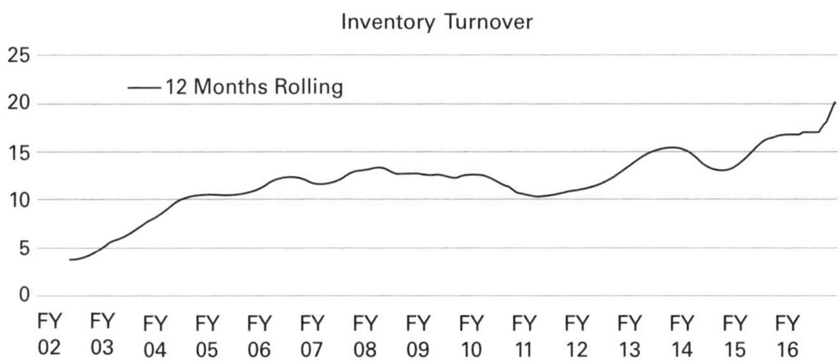

Notes

1 Takt time is the time frame between the start of the manufacturing of a unit and the next. TMHMI calculates it as the ratio between the total available time for production, which is 460 minutes per shift (pauses excluded), and the total volume of trucks that must be produced.

2 Our order book; all the items in the system that still have to be manufactured.

3 Material Requirements Planning is a technique that calculates the requirement of materials and plans production and purchase orders accordingly, taking into consideration demand, market, bill of material, production and purchase Lead Times and available stock. It is seen as a Push model, because it is mainly based on forecast and minimum batch limitations.

4 Minimum batch identifies the smallest quantity of product that can be manufactured or ordered while meeting cost targets.

5 The First In First Out logic involves taking the oldest component. This ensures that the materials do not remain in the warehouses for long periods of time in order to avoid quality issues due to ageing.

6 This is the abbreviation of the word *Yoko-tenkai*. *Yoko* means 'sideways, alongside' and *Tenkai* means 'development, sharing'. It represents the concept of transferring effective solutions from one process to another.

07

2S (*Seiri* and *Seiton*)

This chapter focuses on what are considered to be the foundations of TPS. Any potential idea for improvement or activity related to the implementation of TPS cannot exclude in any way the application of 2S.

2S (Figure 7.1), made of *Seiri* and *Seiton*, is the first two activities of the larger 5S concept.

2S is the starting point for Toyota, before moving to the more complex 5S. Let's first see what is the meaning of 2S:

- *Seiri* (**separate**) means: 'separate what is really needed from what is superfluous and not useful in the workplace'.
- *Seiton* (**reorganize**) means: 'define an order of tools and equipment so that they are immediately available when required'. The goal is to reduce to a minimum the amount of items without causing production delays or interruptions.

As often is the case with Toyota, the application of TPS requires a substantial cultural shift of those involved; the consolidation of a new concept or method needs a gradual implementation. The risk is that, as soon as management eases its control over the application of the new instructions, things can revert back to their previous stage. This also applies to 5S, which might

Figure 7.1 2S in pictures

Figure 7.2 Definition of the 5S

- 整理 SEIRI - *Sort*
- 整頓 SEITON - *Set in Order*
- 清掃 SEISO - *Shine*
- 清潔 SEIKETSU - *Standardize*
- 躾 SEITSUKE - *Sustain*

look easy to understand, but is extremely hard to apply in a systematic and integrated way as a part of corporate culture. With *Seiri* you start separating what is useful and what is not; what is useful is given a specific collocation with *Seiton*; the other 3S are implemented at a later stage, once the first two are consolidated and fully integrated at corporate level. Figure 7.2 shows the meaning of each S.

Usually, the first stage is to focus on the workstations along the production lines: often there are several tools and utensils placed on a nearby table, most of which are not needed for the specific operation that has to be performed. I remember the first actions taken by TMHMI on the matter; originally, each station had a toolbox with a complete set of utensils and tools. Looking for the right tool took a considerable amount of time, but for the operators that toolbox affirmed their identity as a mechanic, ready to face any issue with a wide array of tools.

This toolbox was soon replaced by a rack holding only the equipment needed for that particular station. This has immediately brought an increase in efficiency, but most importantly a reduction in mistakes. It took time and training to accept this modification, but once it became part of the daily routine, we reached a point of no return, because in the long term even the operator acknowledges the benefits created. Nevertheless, a crucial factor to ensure that 2S is fully implemented and followed is the strict control and supervision by management, which must also ensure that all the areas are thoroughly cleaned, regardless of the operations performed in them (this includes the areas that generate more dirt more easily, departments such as welding and mechanical processing). In addition, TMHMI set up an assessment procedure, periodically carried out by management, which evaluates several elements based on the requirements of 2S. Each area receives a rating based on how regularly proper activities are performed; total results are displayed in a radar chart, which highlights the gaps between previously stated targets and actual results.

Beginning from the stations (Figures 7.3 and 7.4), 2S spread throughout the whole plant, becoming a vital element in every working area. The implementation of 2S was carried out following the steps shown in Figure 7.5.

Figure 7.3 Typical early workstation

Figure 7.4 Workstation after the implementation of 2S

Figure 7.5 The steps required for the application of 2S

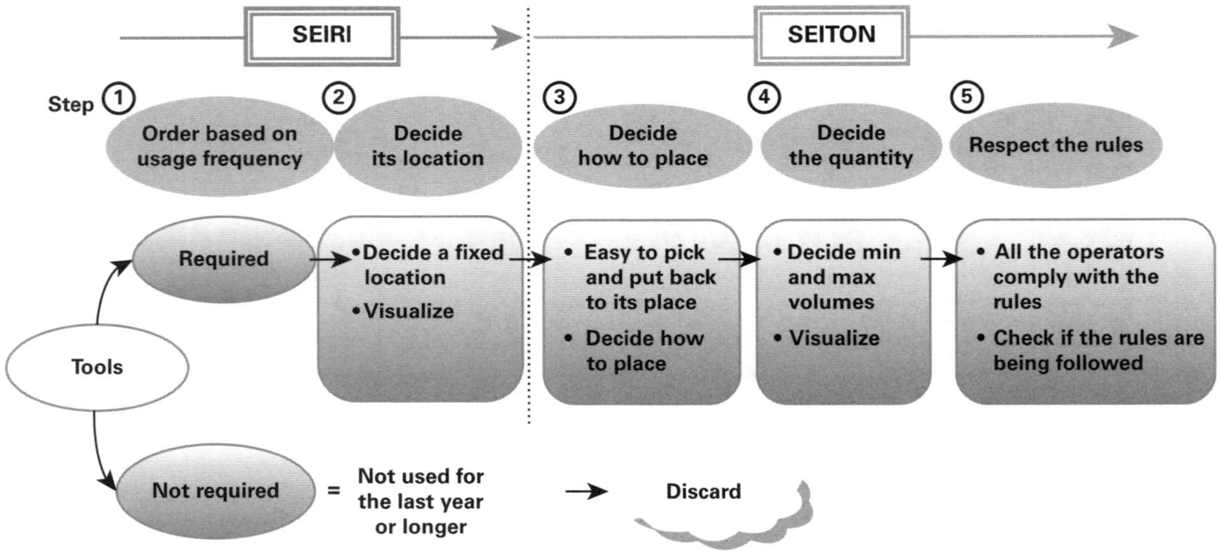

One of the most important results achieved with 2S is the collocation of each item in a dedicated space; the rule that drives this operation is: 'A place for everything, and everything in its place'.

For example, when dealing with the components we created a very clear plan, stating where to place the containers, the areas from which to pick the parts and those that were interdicted for safety reasons (ie because the fire extinguishers were stored there, or because the areas are a clear path towards the emergency exits). In time, while we were applying 2S deeper and deeper in the company, we realized that the more detailed we were while allocating tools or equipment, the easier it was to notice things that were not in their proper place. Basically, order draws order: a tidy area makes it easier, almost intuitive, to spot things that are in the wrong place.

08
The system for feeding materials to the lines (logistics)

One-piece flow required a massive development of the logistics system, in particular for feeding the components to the production lines. Until 2007, with lower volumes and a smaller product line than today's, the materials were placed by the side of the line: the operator would collect directly from the containers the parts required for the assembly of each specific model (Figure 8.1).

When volumes and model range increased, this line feeding approach started to show several limitations due to:

1. the space required next to the lines to stock the materials;
2. long pick-up times by the operators (*Muda Walking*);[1]
3. potential mistakes when selecting the required components (a product is made of several parts, some of which are very similar to each other, but not identical);
4. difficulties faced by logistics when they had to feed the containers by the lines, in particular those who were closer to the production lines;
5. safety issues caused by the lack of a clear separation between the working area and equipment handling.

Feeding needed to be revolutionized; for this purpose, we decided to introduce the concept of 'kit and supermarket', created by Toyota. This feeding model was implemented by Taiichi Ohno, one of the main founders of TPS, who was impressed by American supermarkets during a trip to the US. The basic principle is that in large-scale retail outlets, the customers can find everything on the shelves, without the need to order anything or wait for the product to be retrieved by an assistant (for example, for a can of Coke that

Figure 8.1 Components layout by the production lines in 2007

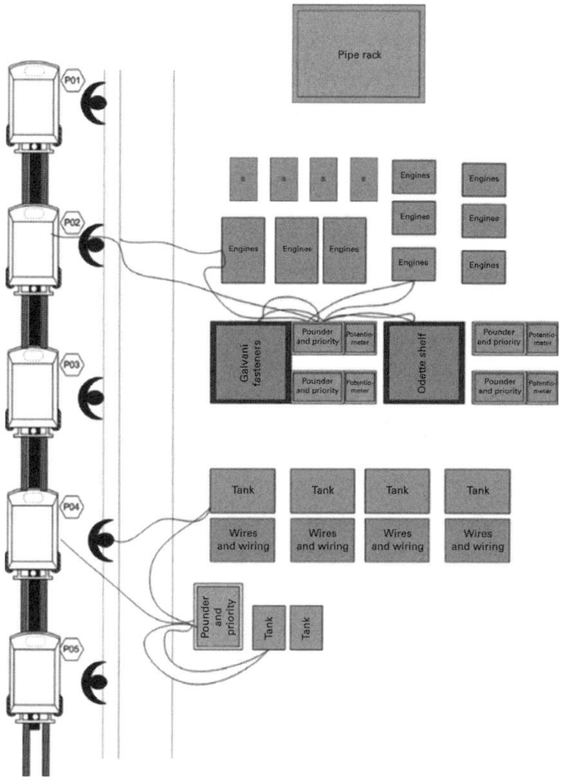

they want to buy). One can pick up anything just moving around with a shopping cart. The application of the same principle in a factory means setting up a similar system of shelves, where all the required parts are displayed and can be easily found and retrieved for the production of the item ordered by the customer.

Such a simple, linear system, while it is taken for granted in the retail sector, has caused a drastic transformation in manufacturing. In our case, once we realized that it was imperative to change the feeding system of the production lines, we moved over a single weekend all the parts stored next to the lines and placed them on shelves built at operator height, divided by aisles and gathered in a dedicated area, which is called the *supermarket* (Figures 8.2, 8.3 and 8.4).

We also assembled small trolleys for the transport of the parts, and a pick list for each order that the logistics operators must follow to retrieve the parts and create the kit. At the end of the route, the logistics operator

Figure 8.2 Production lines in 2007

Figure 8.3 Production lines today

would consign the trolley, which we called the kart, containing all the required parts to complete one or more stations along the line. The kart would follow various stations of the line, moving in sync with the forklift being assembled. At the last station of the line, the kart was placed in a dedicated area, where it was picked up again by the logistics operator to

Figure 8.4 Production lines layout in 2012

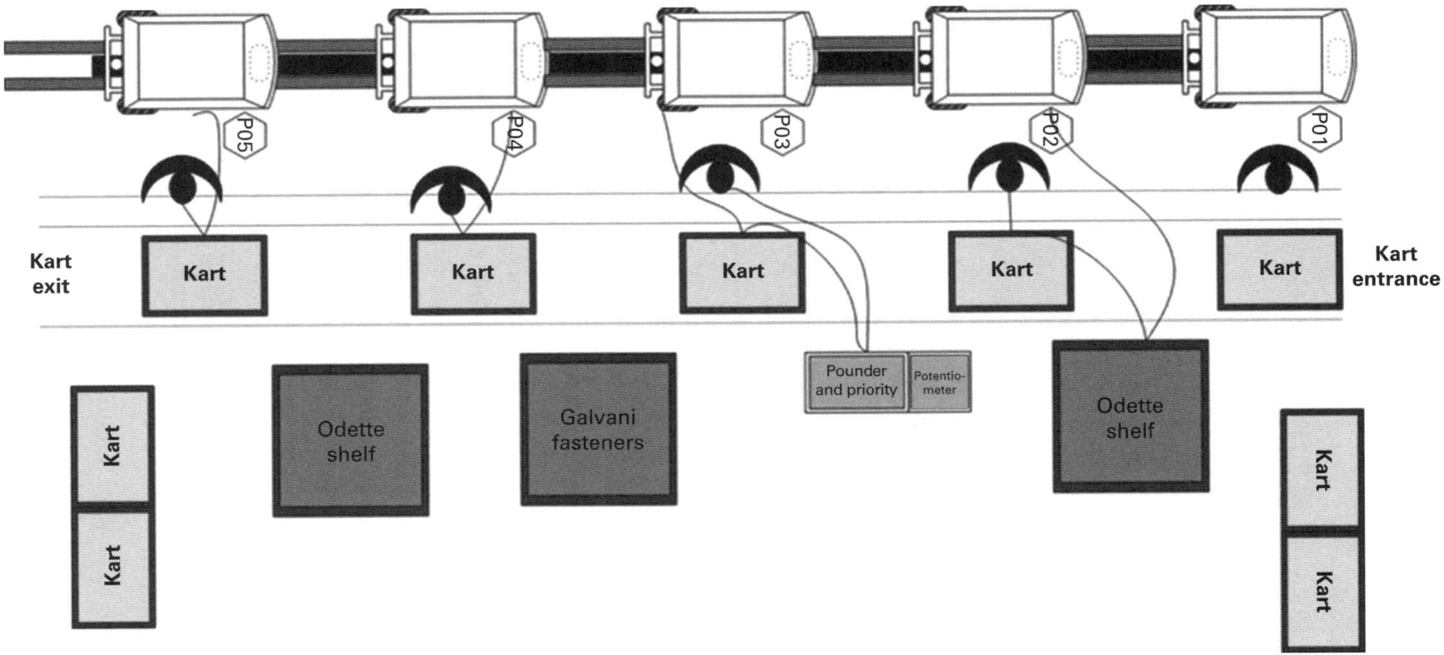

create a new kit. This system was called 'petal', because the route of the kart looked like a flower petal. The whole assembly process could be performed using three separate karts, each one containing the parts required for a single section.

The introduction of the supermarket and kits was crucial to increase the production capacity of the line, improve quality and allow for maximum assembly flexibility. The following steps included the replacement of the pick sheet with a barcode scanner which, based on the bill of materials, according to the model to be assembled, displays the part to be picked up and the quantity required, further reducing the risk of errors. Thus, the line operator must focus exclusively on the assembly of the parts that are supplied, reducing movements to a minimum. If at the end of the cycle there are parts left over on the trolley, it is a clear signal that something has not been properly assembled.

The kit feeding system that we adopted offers several other advantages:

- high flexibility with the assembly of several different models on the same line, in compliance with the one-piece flow principle;
- elimination of setup times for product feeding by logistics;
- clearly visible production lines, clear from all the clutter caused by the materials stored nearby, which contributes to order and tidiness, fundamental dictates of 5S;[2]
- improvements in safety, because the operators have more space available;
- more freedom to install additional equipment along the line.

Once we assessed the benefits that this new system implied, we decided to apply the same principle to the pre-assembly areas, where we started supplying the parts to the operators with plastic boxes; this allowed us to free a considerable amount of space near the assembly lines (Figure 8.5).

The application of *Junjo* has hugely contributed to the improvement of line feeding and to logistics in general: as explained in Chapter 6, the production sequence is finalized seven days in advance, and every forklift is identified by its sequence number and the colour that indicates the day of the week. Using this information, logistics now can identify the material just by using the sequence number, unique for each item that must be manufactured.

At the same time we required the supply chain, which has adopted the *Junjo* supply system, to send and identify each single part with its sequence

Figure 8.5 Assembly area subgroup, before and after

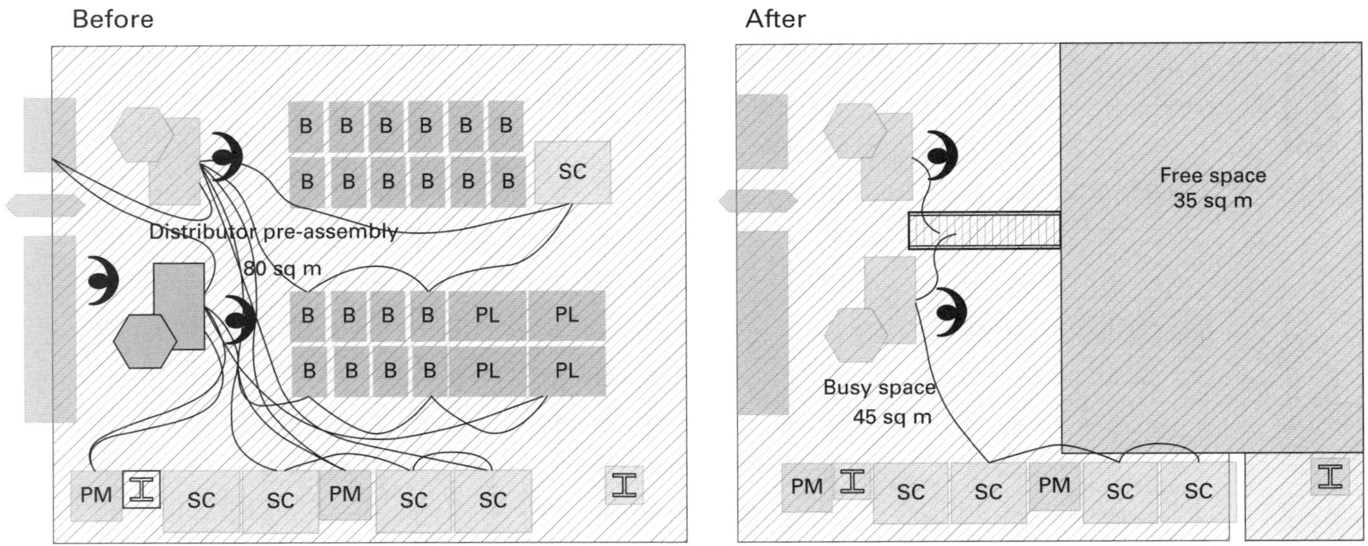

Figure 8.6 Procedure implemented in the *Junjo* area

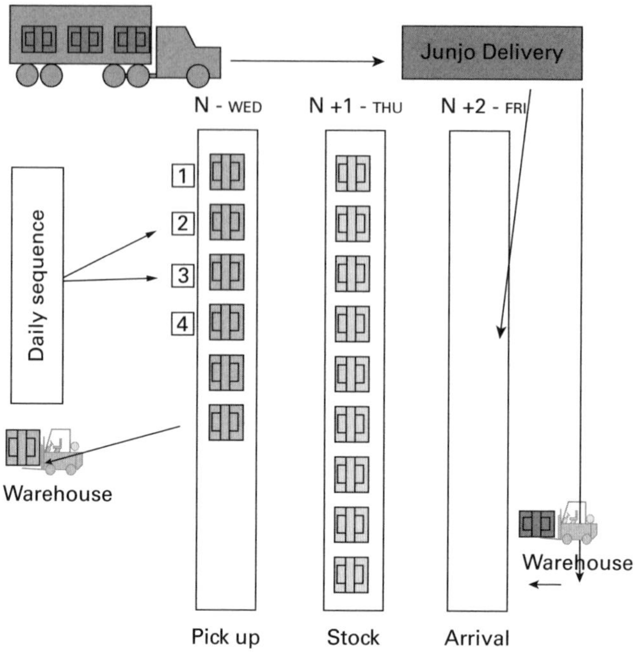

number (Figure 8.6). The use of a card not only drastically simplifies the identification of the materials, but also allows any missing parts to be identified in the sequences that have been planned. The ID card contains the whole code and its barcode, so that inventory is carried out automatically.

The frame was the first part that underwent the *Junjo* treatment. Several models with differing load capacities require specific frames for each range, which means having to manage several codes. Nowadays, the frames are delivered daily by the supplier, unpainted; they are loaded in the truck, each with its own *Junjo* card identifying sequence, code and other information. The operator who is in charge of unloading picks the parts and places them in sequence on a grid showing the line production order. The grid is numbered from 1 to the maximum daily capacity, and divided by days of production. Basically, each frame is identified by a sequential number, written on a card that is placed on the frame itself, also indicating its position on the grid. Due to the fact that the frames are painted in-house, we decided to hold to a maximum stock of three days, during which the painting department retrieves the frames to be painted, always and rigorously following the established sequence.

Once the frames are placed on the numbered grid, their picking order is simplified (Figure 8.7) because it is no longer necessary to identify the part code: the sequence number is perfectly adequate for the scope. This method allowed us to:

- simplify logistics activities (saving time and improving quality);
- limit the maximum stock of an expensive and cumbersome item such as the frame to three days of production (Figure 8.8);
- achieve FIFO;
- immediately spot any missing part.

The application of *Junjo* to the supply chain was extended, including parts that met the following requirements:

- high number of codes due to a large amount of configurations;

Figure 8.7 *Junjo* area where the frames are stocked

Figure 8.8 Days covered with the existing stock after the introduction of *Junjo*

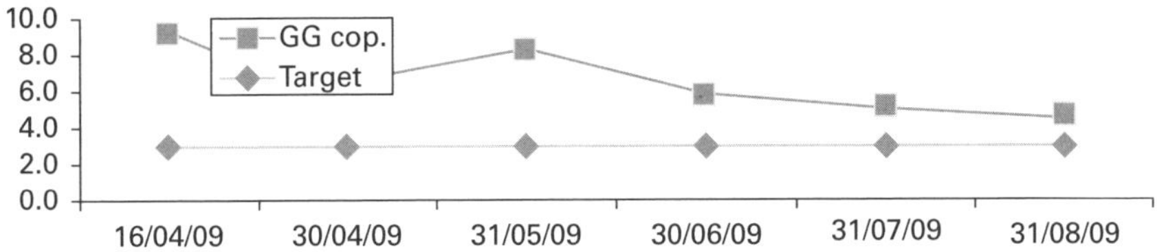

- high cost/volume of the part;
- locally supplied.

As far as the mast is concerned, meaning the part that lifts the weight in a counterbalanced forklift, it is manufactured in a plant belonging to the group where the same methods are applied, which allows us to have a maximum coverage of 0.5 days of production; moreover, the mast needed for about four hours of sequential production are left in the trailer, positioned in the outdoor square/warehouse of the plant, saving us the time we needed to unload and store the cargo. The operator picks the mast needed for the sequence that has to be carried out by the production line directly from the trailer (Figure 8.9).

The last factor related to *Junjo* has to do with transport. Daily deliveries are arguably the best suited system for the criteria set by *Junjo*. We assessed our suppliers' capability to perform daily deliveries; for those who couldn't because of volume or quantity issues, we set up a milk run[3] system, a daily, optimized route to collect items from several suppliers.

The introduction of *Junjo* and its configuration as the logistics pacemaker allowed us to apply other key principles of Toyota's philosophy, among which is the *Jundate*, an area for setting up the components, which we will discuss later.

One of the fundamental principles of Toyota's approach is that the production line has to be 'fed', letting the operators focus mainly on their direct production activities.

There is an anecdote on the subject about the various *sensei*[4] who have supported us in the development of TPS. Often, during their visits, they spent time silently observing the production lines. They would place themselves at the beginning of the line and observe it for several minutes, without uttering a single word. At the beginning I could not understand their behaviour; I thought that it amounted to a massive waste of time. However, my increasing experience with

Figure 8.9 *Junjo* carriages

TPS and several visits to Japan eventually made me understand the great importance of this activity. Watching the lines and the operators at work means assessing the effectiveness and organization of the whole company. One of the aspects that impressed me in the Toyota plants, particularly in Japan, is the pace at which the operators move. Observing their motions clearly shows the sequence of activities they carry out, the harmony of their implementation and the precision of each step. There are no indecisions or pauses: everything is organized so that the flow is constant and uninterrupted. Achieving such a result implies that each movement of every operator has been studied so that they can find the tools and the components in the best possible location: everything is in its right place. This means total quality of the components and the ability of logistics to supply the lines while complying with required volumes and timelines. I discovered this world during one of my first training experiences in Japan, which I spent exclusively on the production lines.

I felt like I was in the middle of an orchestra, where each instrument played the right tune at the right time, contributing to the general harmony of the whole structure. In order to achieve such a remarkable result, every single detail had been studied, improved over time and clearly defined.

'There is no genius in Toyota; we just do what we think is right, trying to improve something every day, step by step. Fact is, if small improvements add up for seventy years, they become a revolution.' This philosophy is what made it possible, over time, to implement a system capable of creating such harmony, by paying the utmost attention to every single detail, from the assembly sequence by the workstations to the feeding system of the production lines.

The karts that carried the kits prepared at the supermarket were eventually replaced by roller conveyors installed by the production line (Figure 8.10). The components, prepared in the supermarket to fulfil the needs of each station, are put into containers that are then placed on the

Figure 8.10 Roller conveyors used to feed components in *Junjo*

conveyors. This system further reduced operator movements, since it is no longer necessary to move the kart from one station to the next.

Besides the supermarket, other areas have been set up to split more voluminous components. These areas, called *Jundate*, are where the parts are picked based on the programmed production sequence and placed on specific carts that are subsequently towed. On every trip of this small train (Figure 8.11), the materials required for the assembly of three trucks are brought to the lines; basically, the 'train' supplies the lines every three Takts.

With the introduction of *Junjo*, each forklift fed into the production process is identified with a number and colour. In order to simplify its identification, an A3 summary sheet was developed and is attached to each truck during the first production stage. This sheet, called the infomatrix (Figure 8.12), reports all the main information that allows the production process to proceed.

The upper area of the sheet shows information such as production order, model, customer and stated delivery date: all the main data that the operator will use to check if the components prepared by logistics match the production number.

In the lower area of the infomatrix we can find individual sections that show the options and configurations required by the customer, namely:

- cab;
- seat;

Figure 8.11 Feeding 'train' in *Junjo*

Figure 8.12 Infomatrix

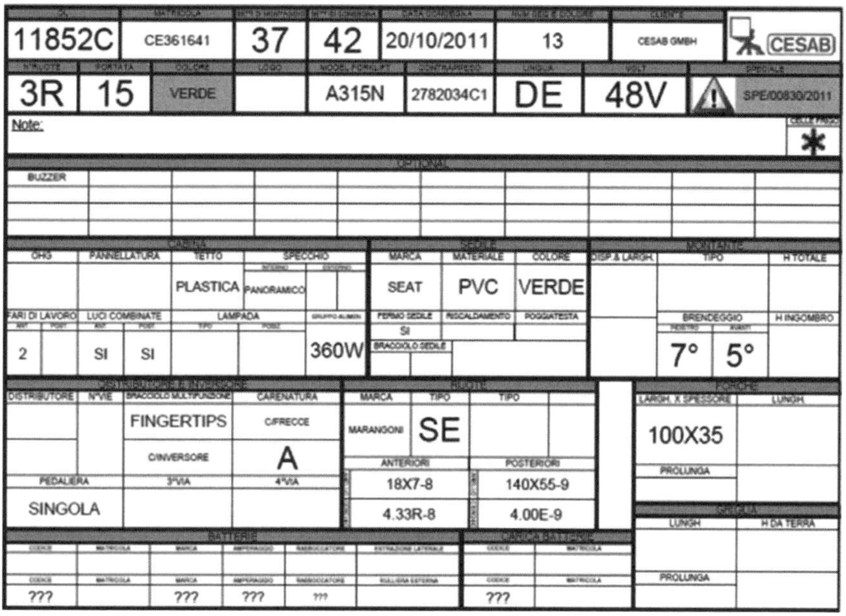

- mast;
- distributor and reverser;
- wheels;
- forks;
- batteries;
- battery charger;
- backrest.

The goal is to simplify the understanding of the configuration required by the customer, and to verify if there are any discrepancies with the components that have been prepared.

The evolution of these systems meant that logistics now play a fundamental role, upon which is based a good share of the company performance: its reliability and precision are indispensable to reach productivity targets, quality and OTD. Today's frequencies imply 12,000 to 13,000 movements per day: 1 per cent of wrong parts or quantities can cause more than 120 mistakes, which would have a serious impact on the plant performance. Moreover, such a complex system required total standardization of the activities carried

Figure 8.13 Logistics flows and warehouses within TMHMI

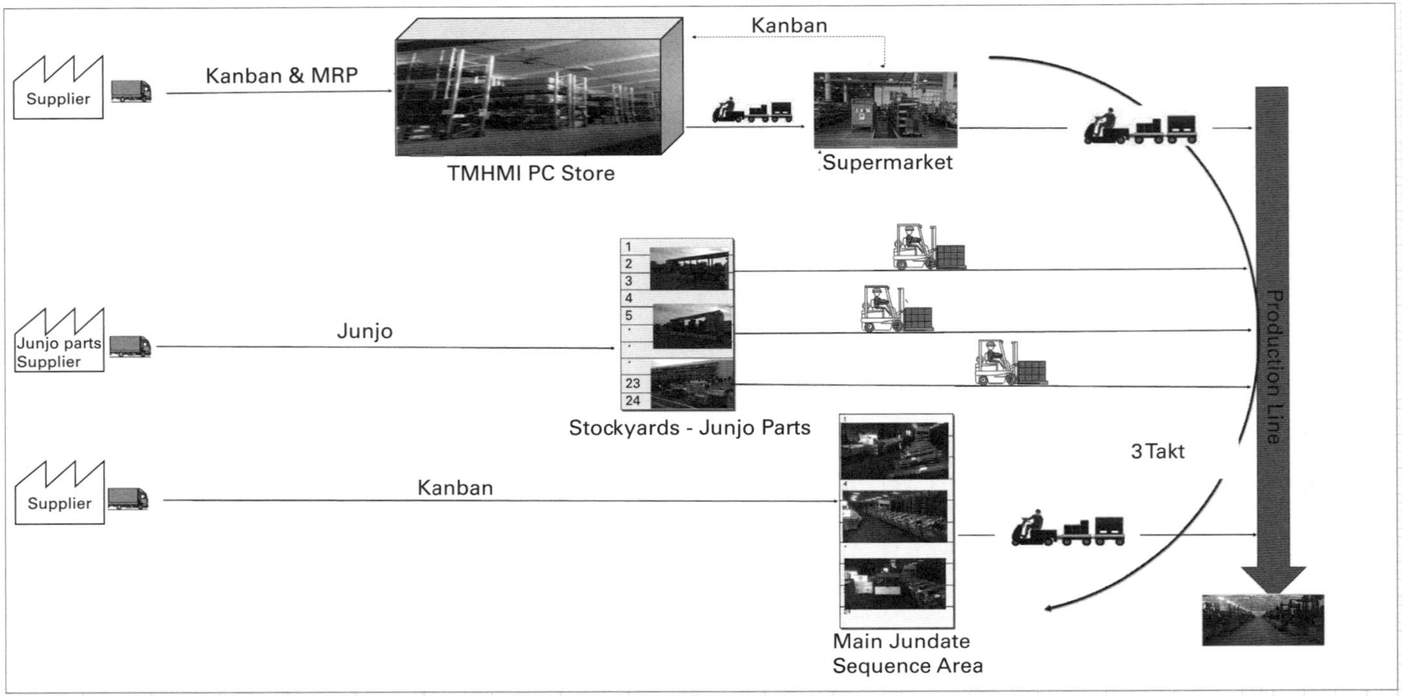

out by the operator, which in turn required a procedure where everything is organized to the finest detail: there are working cycles with strict timelines to comply with; picking routes have been studied and optimized; electronic devices have been adopted so that the operator not only can identify the correct part, but can also record the relevant warehouse inventory variation through the WMS[5] (warehouse management system).

The different procurement and feeding system of the production lines had a considerable impact on the internal layouts and on the supply structure that was being used. We began to rely on towing tractors, to transport the materials from the storage areas to the production line. Figure 8.13 shows the current logistics flow of TMHMI and the main handling processes of the materials.

Notes

1 TPS sees any motions by the operators during the production stage as waste.
2 5S, representing order and cleanliness, is the basis of TPS and is described in Chapter 7.
3 Logistics model that sets a collection route from several suppliers. With a single journey, we can retrieve small quantities of materials from each site, optimizing the truck load and the volumes reaching our warehouse.
4 Person born before another; usually it means 'master' or 'expert'.
5 WMS is a software that is usually integrated with the managerial IT side of the company; it oversees the movements of materials in the warehouse and processes transitions, including shipping, receiving, reorganization and collection. It is used to map the materials and allows for the optimization of the operators' picking routes and the management of the locations based on rotation (ie lower shelves near the exit for high-rotation materials, upper shelves far from the exit for parts with lower turnover).

09 The organization within Toyota

In order to support the cultural shift required to apply TPS and spread it through the company successfully, we needed to make changes to the organization, to roles and activities managed by its members; we had to give a strategic role to the group leaders (GLs).

Toyota faced heavy confrontations with Japanese labour unions from the end of the '60s to the beginning of the '70s. The clashes between management and unions caused strikes and production freezes. Such mutual lack of trust highlighted the communication difficulties and caused several modifications, from the creation of open-space offices (including top management desks) to the reassessment of the duties of internal figures such as the GL. This role is so relevant for the company (Figure 9.1) that the education of GL trainers, basically the training of those who will in turn train the GLs, is made directly by the parent company in Japan, so that the whole group maintains a common imprint.

A specific, 15-day course held by HR illustrates the activities and characteristics of the employees assigned to this role.

Some of the GL main duties involve:

- Daily management: establishing and managing the KPIs of their department.
- Operators management: sharing and referring to HR training plans, bonuses and concerns about employees.
- *Hoshin Kanri*: working to achieve the corporate goals defined by management.
- Organization and development: contributing to the development of the company and of its human resources, creating an environment where people collaborate and can express their full potential.

A GL manages a department or a production line; they usually coordinate no more than 70–80 people and are supported by the team leaders (TLs) for their

Figure 9.1 GL functions in Toyota

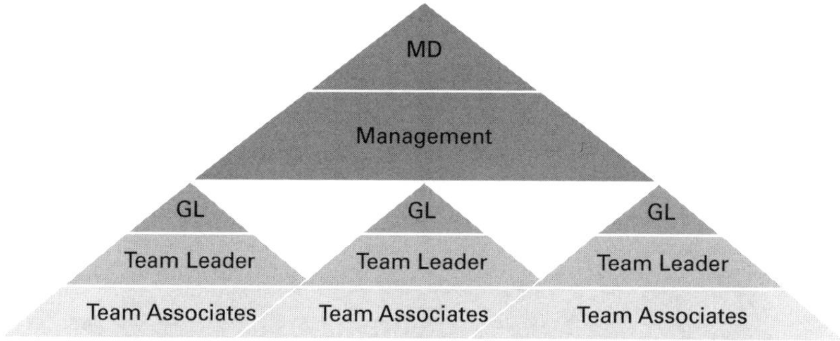

operational duties. Aside from the activities listed above, GLs have a double role in the company: they are a link between management and operators, enabling communication, information and company strategies. They also perform the same job the other way around, reporting to management the situation of the department and any possible tension that might need to be dealt with. Often the GL role is assigned to employees who have a qualified experience in the production area; this choice affirms and strengthens their leadership over the operators.

Team leaders manage up to 10 people and assist the operators in their production duties. Support is given directly, helping the operators in the execution of their work when required. The TL is called to the station in need of help through the *Andon* board, which is installed in all the plants of the group. This board was originally developed to display the advancement stage of production and to show any anomalies, so that they could be solved immediately, complying with one of the fundamental elements of *Jidoka*. The board has evolved considerably over time, gradually becoming more technological and advanced, extremely flexible and capable of recording and displaying several kinds of information. Evolution notwithstanding, the basic information has remained unchanged. Therefore, every *Andon* board (Figure 9.2) in every Toyota plant displays:

- workstations;
- daily production target;
- the target for that specific moment in time;
- the volumes produced so far;
- downtimes occurred during the day.

Figure 9.2 *Andon* display in TMHMI

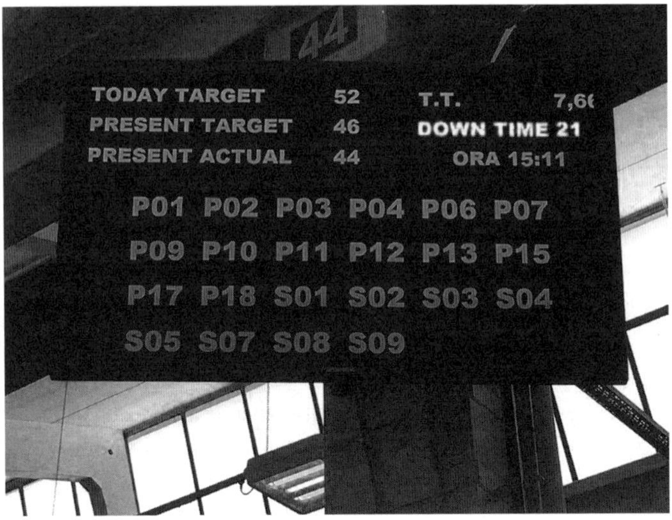

In Japanese production lines, the *Andon* also shows the overtime that is planned for the day: there is an agreement with the unions that allows the company to require up to two hours of overtime every day, if this decision is notified by 3pm the same day.

If the operator encounters issues during the production stage, they call the TL with a button that is located by each station. The *Andon* immediately displays the call, the workstation on the board is highlighted in yellow, while an audio signal calls for attention and reduces response time. The team leader assesses the issue with the operator, trying to solve it within the takt time. If this cannot be done and the truck goes out of phase, the TL can decide whether to stop the production line, in which case the workstation number on the *Andon* turns red and a downtime counter starts, or to let the truck proceed to the next stage, indicating the problems that will eventually be rectified in the dedicated service area.

In addition to the support given to the operators, team leaders manage other activities in collaboration with them; one of the most important is focused on improvement: the operators refer to team leaders to discuss potential issues and suggest possible solutions.

10
Standardization

One of the main factors of TPS is standardization, a term that has many connotations but that in Toyota can be summarized with a quote by the president, Mr Toyoda: 'Our competitors have excellent people in mediocre processes, while we have normal people in excellent processes: victory can be only ours.' Standardization leads to excellence, and such excellence becomes an asset of the company, available to everyone.

Standardization is therefore the application of what people successfully managed to improve through their personal experience into processes and information for the entire group. There are other companies that rely on the skills and capabilities of a limited number of people; when these people are not working, the system cannot ensure the same quality and performance standards.

Many enterprises can boast highly capable and experienced operators in their workforce, while it is much more difficult to find places where such capabilities and experience are applied to methods and processes. I remember well what happened when we began a deep improvement effort of the product quality by adopting the 4M approach[1] (Man, Machine, Method, Material) to the production lines. With this system, each quality issue is assessed and classified based on the four areas above. The issue is then analysed in depth until the real source of the problem is identified. Every time we applied this method, the cause turned out to be Man: it looked like the operator was always the culprit in each occasion. The results looked suspicious; I doubted that we had such a large number of 'useless' employees, so I personally verified the situation on the lines and found the real truth: the instruments and the information we made available were totally inadequate for them to assemble the trucks the way we expected. I realized as such that those working conditions were not excellent as they should have been, and that they required excellent people to be carried out successfully.

This approach is one of the cultural differences that I've noticed between Japan and Italy (but not limited to these countries). In Italy the attitude is often to find who is responsible, rather than assessing the system as a whole. During a training course I attended at *Hazu Academy*, the Toyota training centre in Japan, I had the chance to spend two weeks with the company's top management. The HR vice-president of the group commented on the different approaches of our countries when it came to problem-solving: in Europe we have the tendency to work on *who is wrong*, while Toyota prefers to work on *what is wrong*. In our case, the company had to assess whether its employees were given all the instruments they needed to perform at the best of their abilities. We are not only referring to 'technology' tools: the concept also applies to the working environment and to the interpersonal relationships among people performing different functions at different levels. This factor is part of teamwork, one of the Toyota values, which we will discuss in the following chapters.

Standardization requires the implementation of critical elements, among which:

- takt time;
- operating sequences;
- stock levelling among processes.

Takt time is a key concept within Toyota: it represents the frequency at which the customer requires the product, and it is the ratio between net daily working hours and the quantity of required products. Takt time must rule over all the processes in a department/business; ideally it should be the same for everyone, because in a well-balanced factory, each process manufactures the same volume of products in the same period. This principle requires an unequivocal vision of the factory, which adds one unit per takt. In TMHMI, takt time is allocated to the assembly lines on the basis of customer requests; as a consequence, all the processes are balanced with the same takt. The only exception concerns the painting department, which works over two shifts rather than one to fulfil the orders. The allocation of a frequency, or a number of products delivered in a time slot, implies the organization of a work sequence and the analysis of the time required for each and every operation, so that the operator has a working cycle that is compatible with the allocated takt time. Although in the Japanese factories of the group one can see an amazing speed of execution by the operators, the cycle given to each station must not cause excessive fatigue on the workers, in order to avoid issues of quality and safety. Oversaturation of the operators is one of three kinds of waste that are defined by the TPS philosophy, which identifies it as *Muri* (the others are *Muda* and *Mura*).[2] With standardization, each activity is singled out and analysed, so that those risks can be avoided. The first step towards the standardization process in

TMHMI was the creation of the working sequence, composed of all the documents meant to the operators, indicating all the procedural details for each workstation.

We developed two main documents applying the standards set by Toyota: the WES (Work Element Sheet) and the WSS (Work Sequence Sheet). The WES is used to describe the assembly sequences for each station, the model they are related to, if they are STD operations or specific for a particular model, and the reference documents that can be consulted. The WSS includes further details, showing the procedure of each task with pictures and drawings (Figure 10.1).

Quality and standardization had great success in Japan thanks to W. Edwards Deming; the wheel that illustrates its main points is still used by Toyota in many activities.[3] Deming's wheel is split into four parts (Figure 10.2):

1 **Plan:** identify the problem and plan the corrective actions to solve it.
2 **Do:** carry out the plan.
3 **Check:** check whether the results are in line with the expectations established in the Plan stage.
4 **Act:** implement a correcting action if the results are not in line with the target; otherwise, standardize and consolidate.

One of the main differences that I noticed between Japanese and Italian culture is that the Japanese consider the Plan stage as crucial, where they focus most of their attention, while Italians tend to see Do as the main element; this approach, as illustrated in Figure 10.3, poses a high risk of delays and could not bring the desired results.

Figure 10.1 WES and WSS examples

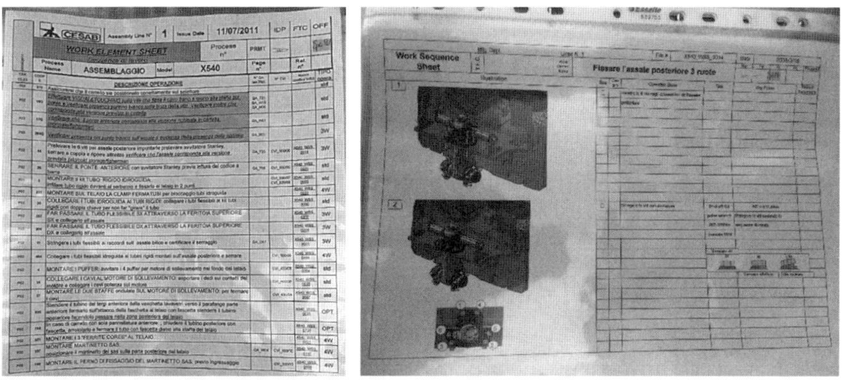

Figure 10.2 PDCA

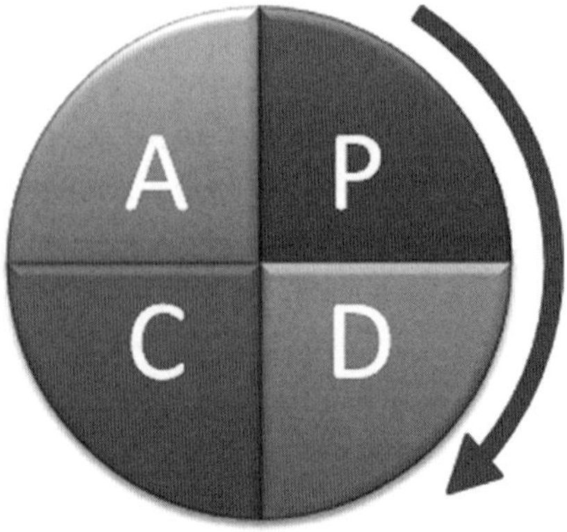

Figure 10.3 Accuracy of planning versus time

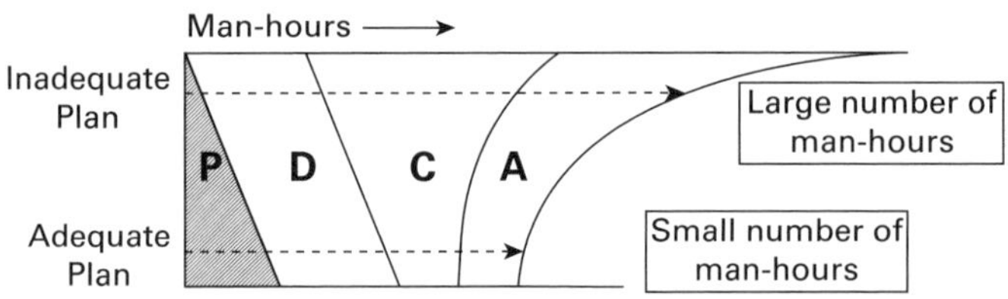

Let's imagine that we want to take a bicycle journey, and we cannot bring the satellite navigator with us. A good Plan approach requires that before leaving we identify all the turns that we need to take at each junction we cross along the way, so that when we reach the Do stage, basically when we start our journey, we can optimize our riding time. If, instead, we neglect the Plan and leave right away, we would be forced to stop at every intersection and look at the map to find the right turn. I am positive that by using the first approach we can reach our destination much faster. It is also very important not to overlook the Check and Act stages: checking the results allows KPIs and targets to be identified during the Plan stage, so that we can evaluate whether the project and the activities are producing the results we want. It is a useful tool for implementing the right corrections and, moreover, for setting a standard available to the entire company (Act stage).

Buffers, areas where a set number of units is placed between processes, are part of the standardization process (Figure 10.4). Each unit is clearly visible, in order to spot immediately any anomalies or bottlenecks. TMHMI set up several buffers in their operations; for example, there is a buffer

Figure 10.4 Example of a balancing buffer

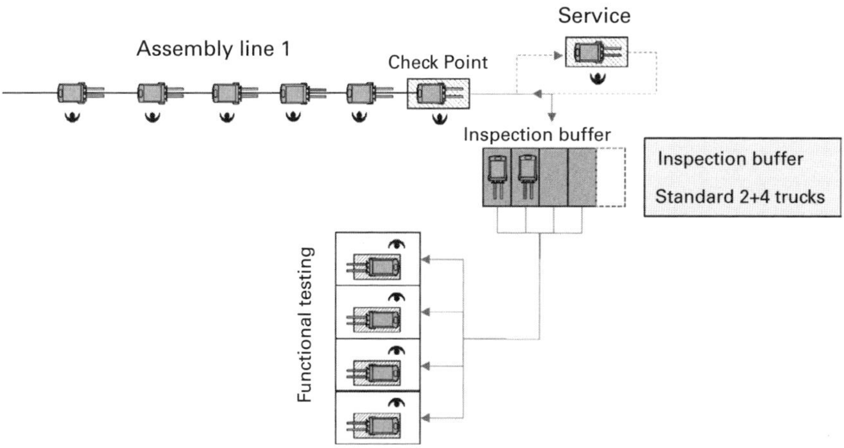

between assembly lines and the first inspection stage. Since both processes have the same takt time, an increase or decrease of the trucks in the buffer shows that the processes are not balanced. For example, too many imperfections detected at the checkpoint at the end of the line can cause the inspection downstream to be delayed, a situation that is clearly shown by the lack of units in the buffer.

Standardization is part of the *Jidoka* pillar of TPS, because it is strictly related to the improvement of quality processes, and it can also identify potential anomalies.

Notes

1 4M, also called Ishikawa diagram or fishbone diagram; developed in Japan in 1943, it displays graphically the analysis of an issue, splitting all potential causes into four categories (Man, Machine, Method, Materials). Each one is then divided by sub-causes, which are studied to identify the main elements that need to be addressed to solve the issue.

2 TPS classifies waste into three different categories. In addition to *Muda*, which is an activity without added value, there is *Muri* (overload, in particular concerning people) and *Mura* (unreasonable, overdoing, which usually causes *Muri*).

3 Liker (2000).

11
Quality

One of the elements that has always been associated with the Toyota brand is quality.

Such quality is not only achieved through excellent inspection and control systems, which are capable of intercepting any quality issues, but it is also the result of procedures and verification systems that are carried out at each step during the development of a new product, and are implemented right until the last inspection along the production lines. It would take an entire book to explain in detail the development model for new products that is used by the group: the procedure is composed of various steps, called design reviews, management meetings where the progress of the project is assessed and eventually approved.

Quality is a must for the group in all corporate functions; paying attention to the smallest detail, inside a framework that manages every stage of the process, can improve the process automatically, making it more efficient and precise. In addition, complying with the philosophy of standardization, we strive for continuous evolution and look for the *Kaizens* that ensure a constant development of the system and guarantee its development. Mistakes are examined and solved immediately, becoming themselves an asset for the company: they are part of the information about company standards, so that they will not be repeated in the future.

Quoting Taiichi Ohno, the founder of TPS, 'the biggest problem of a company is having no problems'.

One of the factors that defines excellence in quality is the method of analysis, a crucial part of problem-solving;[1] this approach is integrated in the model that Toyota calls quality control, defining the methods and the main approaches that must be applied when looking for the solution to a problem. Quality control includes the seven tools used for data analysis:

1 Pareto;

2 cause and effect diagram;

3 check sheets;
4 graphs;
5 scatter diagrams;
6 histogram;
7 control charts.

The *Dantotsu* Project

One of the most significant examples that can illustrate how quality is perceived within the group is related to a project launched in 2007 by Toyota's Quality Department: *Dantotsu*, which means 'to be indisputably the number one on quality'. Three main indicators were chosen for this project, each representing a macro-category:

- **Complete Vehicle Inspection (CVI):** this KPI aims to measure the quality of internal processes.
- **Non-Conformity Report (NCR):** this KPI aims to evaluate the quality of the supply chain.
- **Warranty Claim Paid (WCP):** this KPI measures the costs of servicing the products under warranty, therefore assessing the level of customer satisfaction.

The project involved all the plants worldwide; its goal was to reduce quality costs in three years and achieve the following results:

- CVI: –95 per cent (defects found at the end of the production line);
- NCR: –95 per cent (non-compliance by the suppliers);
- WCP: –75 per cent (cost of warranties).

Each plant was assessed on a monthly basis to check its KPI trends; each year a meeting would analyse the results that were achieved. Toyota's senior advisor, who was in charge of the project, would visit each site four times a year for a week, checking with management which were the activities that needed to be improved and which were correctly performed. Each KPI involved specific activities.

Reduction of CVI (complete vehicle inspection)

CVIs, measured as the ratio between trucks manufactured and faults reported, are assessed at three different checkpoints[2] inside the plant. The first is located at the end of the production line, the second is performed after the first functional tests, while the last one is carried out during the final inspection stage, prior to the declaration of conformity that authorizes the delivery of the truck to the customer (Figure 11.1).

The faults that are detected by the checkpoints are related mainly to parts supplied externally or to assembly issues (mistakes by the operator); rarely they involve the engineering process that defined the development of the product.

A daily meeting called *Asaichi* (Japanese for 'morning market') (Figure 11.2) was organized to manage and improve the performance of CVI; during the *Asaichi*, all the GLs illustrate the problems that were encountered during the previous day of production, using pictures or bringing the faulty component to

Figure 11.1 Layout of the checkpoints on the lines

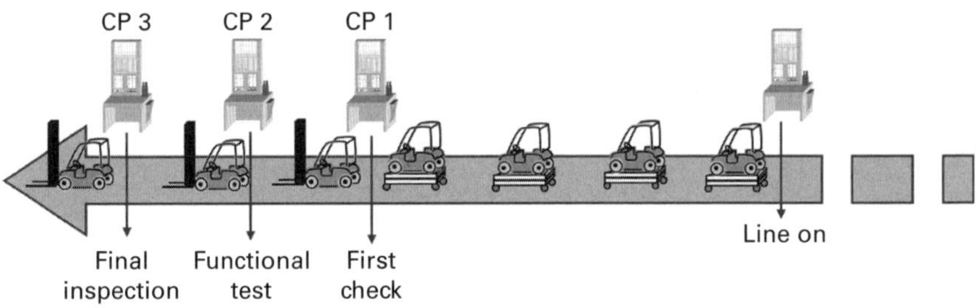

Figure 11.2 *Asaichi* meeting

the meeting. Each problem is studied and a department is given the task to solve it (usually the department that caused the problem in the first place). These meetings are run by the production GL and see the attendance of all of the company's corporate functions; all the graphs and countermeasures are updated daily by the person responsible.

Countermeasures are developed by following the PDCA wheel method; most importantly, for each countermeasure there is a person who is responsible for its implementation and for the solution of that particular problem.[3]

True to the principle of involving the operators, quality management and its KPIs are transferred directly into the plant where the problem has occurred, applying one of Toyota Way's principles, *Genchi Genbutsu* ('go and see with your own eyes'). It wasn't easy to involve all the people not strictly related to production and take them every day to the production areas. The first meetings saw the attendance of the top management of the company, in order to prove that we were committed to this new approach; however, often the meetings did not see the attendance of departments that were not involved with operations. It took perseverance and tenacity to embed this meeting into the company DNA, but the results it managed to achieve gave it credibility over time. Today, years later, all corporate departments attend the *Asaichi*; GLs can manage the meeting perfectly, updating all the indicators, respecting the 15 minutes' allocation and, most importantly, involving all departments to find the countermeasures needed to solve the problems that have been detected.

Such great exposure of all the difficulties that the company must face on a daily basis has contributed to focus the attention of all the departments in order to find the correct countermeasures: each department verifies the list of problems involving them, checking for new reports.

Over time, the meeting has become effective and functional; CVI values have improved noticeably, with two-digit reductions. In order to perform increasingly detailed analysis and spot the real causes with more precision, we introduced the concept of problem-solving and its eight steps, shown in Figure 11.3.

In order to verify whether the analysis of the problem was successful and succeeded in spotting the real causes, and to validate the effectiveness of the countermeasure that was implemented, a graph for each of the main problems shows if the same defect has occurred after the implementation of the countermeasure (Figure 11.4). If this is the case, the process is restarted, reapplying the steps of problem-solving to spot the real issue, find a different countermeasure and monitor its results.

Figure 11.3 Eight steps of quality

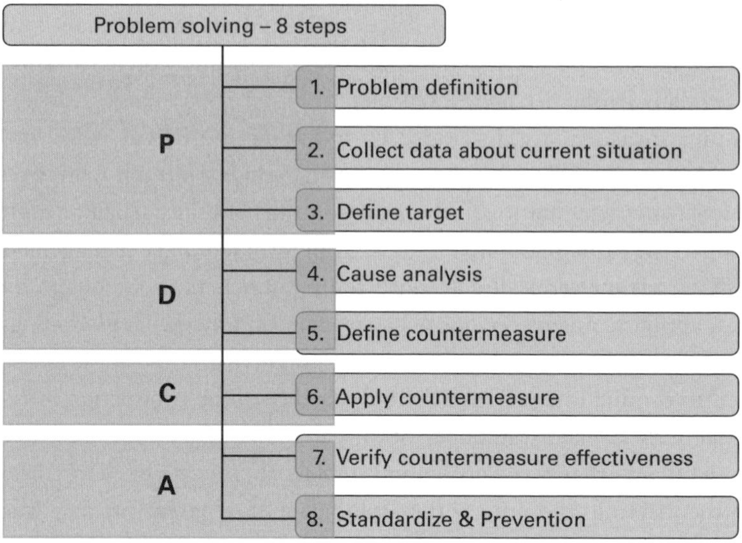

Figure 11.4 An example of problem-solving

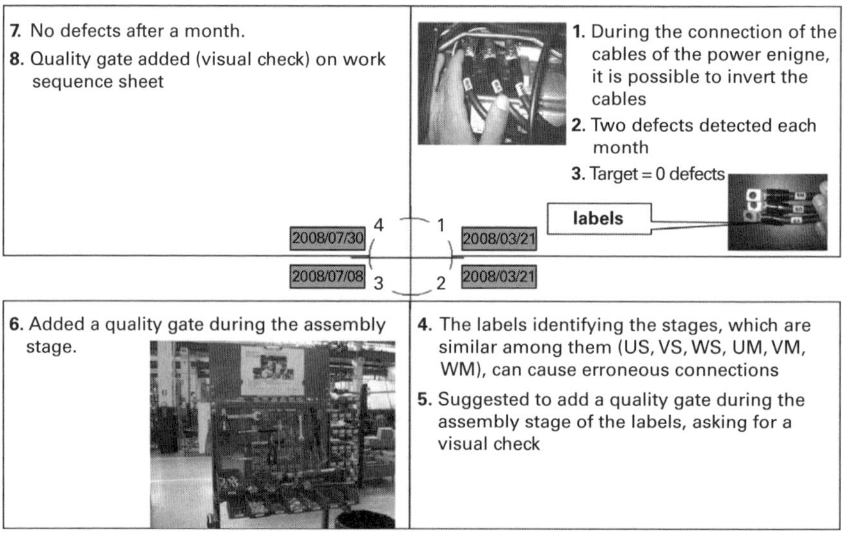

We also made improvements by classifying the defects by area, including the anomalies reported by customers at 'zero operating hours'. Such classification is crucial to improve the checklists used during the final inspection stage: if a customer reports a defect at zero operating hours, the verification of that particular element is added to the lists.

Defects have been classified as follows, according to the suggestions of Toyota's senior advisor:

- type D defects: reported by customers;
- type C defects: reports filed during the final inspection;
- type B defects: reports filed by the final inspections for each single process (ie painting, carpentry, assembly, etc).

Quality gates have been introduced for critical defects, whose responsibility is assigned to the line (Figure 11.5); an illustration of the defect, be it a picture or a drawing, is placed by the station responsible for the issue and reminds the operator to pay particular attention to that particular step.

Just as with any change involving human resources, in the early days we had to show great determination and commitment, so that the tools that we adopted were actually used by the employees. In my experience, I have noticed how each cultural shift goes through a first stage, which I call 'elastic

Figure 11.5 Example of a quality gate

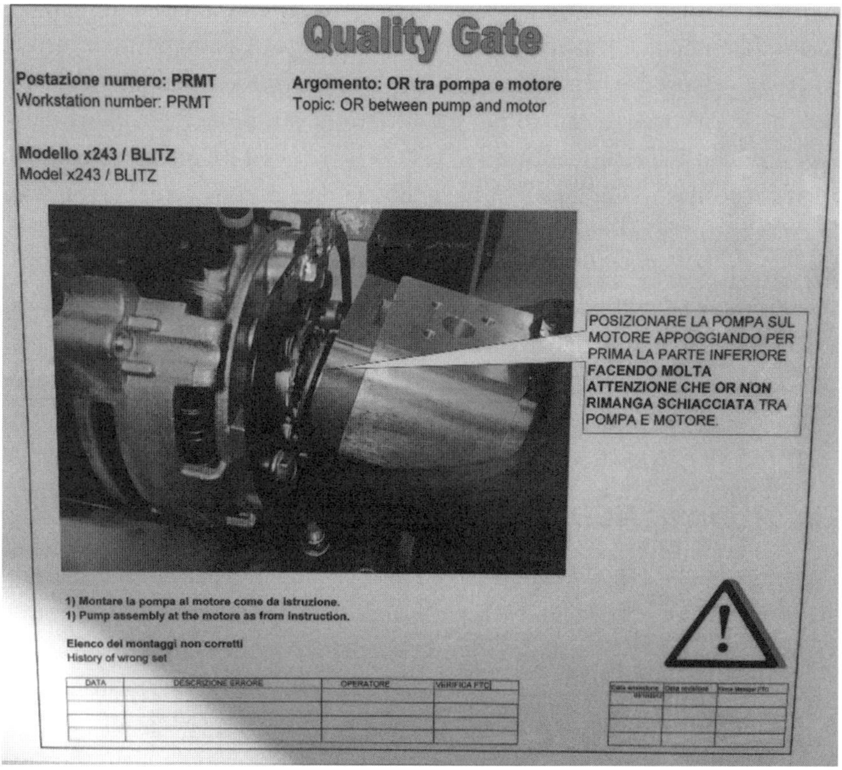

Figure 11.6 CVI performance

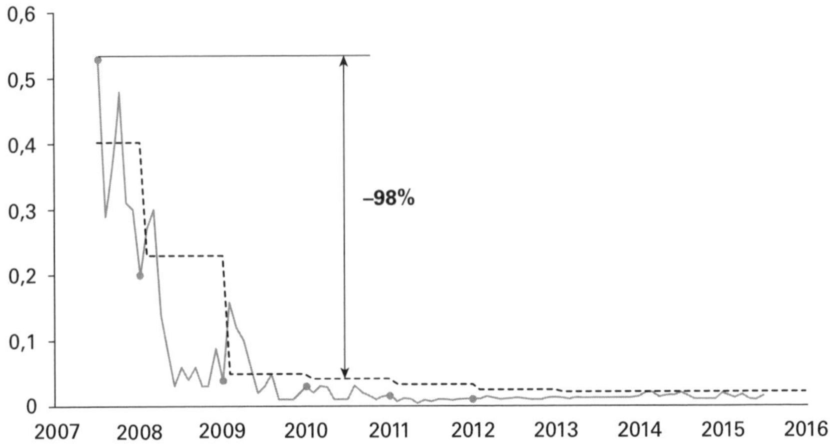

deformation', during which management pressure on the subject must be constant; otherwise, things tend to revert to the old, established procedures.

To achieve what I call a 'plastic deformation', where the change succeeds by consolidating within the culture of resources, management must make a great effort and act systematically towards change. A point of no return is generally achieved when the people involved begin to understand that the tool, be it a process or any other modification, can properly support daily operations: it is performed not because it is required by management, but because those who implement it have fully understood its actual usefulness.

Thanks to the adoption of the activities illustrated above, soon our –95 per cent target, which looked unrealistic at first, was reached and improved. Moreover, as shown by the graph in Figure 11.6, this method allowed us, after some early adjustments and modifications, to maintain our performance levels, even with the introduction of new products and of takt time, events that usually cause a temporary decrease in quality.

Reduction of NCRs (non-conformity report)

On a production site like ours, where the supply chain guarantees a sizeable share of the product value, each performance (be it in terms of quality or customer service level) relies on the involvement of the supply chain itself. Suppliers are carefully selected; R&D and acceptance perform several checks on the parts before approving a new product or a change of supplier. Most importantly, selection and development of the supply chain involve several audits, performed by TMHMI personnel, to verify and assess the consistency of processes, procedures and activities, so that they ensure the quality of the products.

During the *Asaichi* meeting, any quality issues regarding the parts received by the external supply chain are reported and discussed. In such cases, the quality department of TMHMI takes responsibility for the report, informing the supplier with a non-conformity report (NCR). Each NCR triggers a series of actions, such as the inspection of the parts already on site and the exclusion of those that do not meet requirements, so that the line is fed only with compliant parts. The supplier must reply to the NCR in writing, outlining a careful analysis of the issue and countermeasures. Most of the supply chain has been trained on the methods used by Toyota to analyse defects, such as those related to quality control outlined at the beginning of this chapter. Moreover, key suppliers have been trained by TMHMI on Toyota's approach to quality, and they too implemented *Asaichi* meetings in their production sites.

Basically, the same approach and methods developed by TMHMI for the reduction of CVIs have been transferred to the supply chain, with the aim of increasing the conformity of the parts produced. A team of TPS experts supported the people responsible for the quality of supplies, transferring the acquired know-how based on the audits performed on their processes and quality requirements. Our experience as precursors of this method, and the fact that we could demonstrate with data and facts the positive results we achieved, made its diffusion much easier.

The implementation of *Asaichi* in the production sites of the supply chain has simplified and standardized the management of NCRs. Each report, including those received by the customers, activates procedures that involve the sales network, TMHMI as manufacturer of the good and the companies that supply the parts (Figure 11.7).

The adoption of this approach means that every time an NCR is reported by the lines or received from a customer, the referent of the quality team immediately relays the issue to the supplier with a written report, describing in detail the defect that was detected. The supplier will discuss the same issue at its own *Asaichi*, so that adequate countermeasures can be implemented and transmitted to TMHMI. With this system, the person responsible for the quality of TMHMI supplies can reply during the *Asaichi*, illustrating the causes that generated the defect and the countermeasures implemented to avoid the recurrence of the same issue. The supplier itself is often invited to the *Asaichi*, so that it can show personally to the team which countermeasures were applied to decrease NCRs.

Figure 11.7 Flow of quality reports from the customer to the supplier

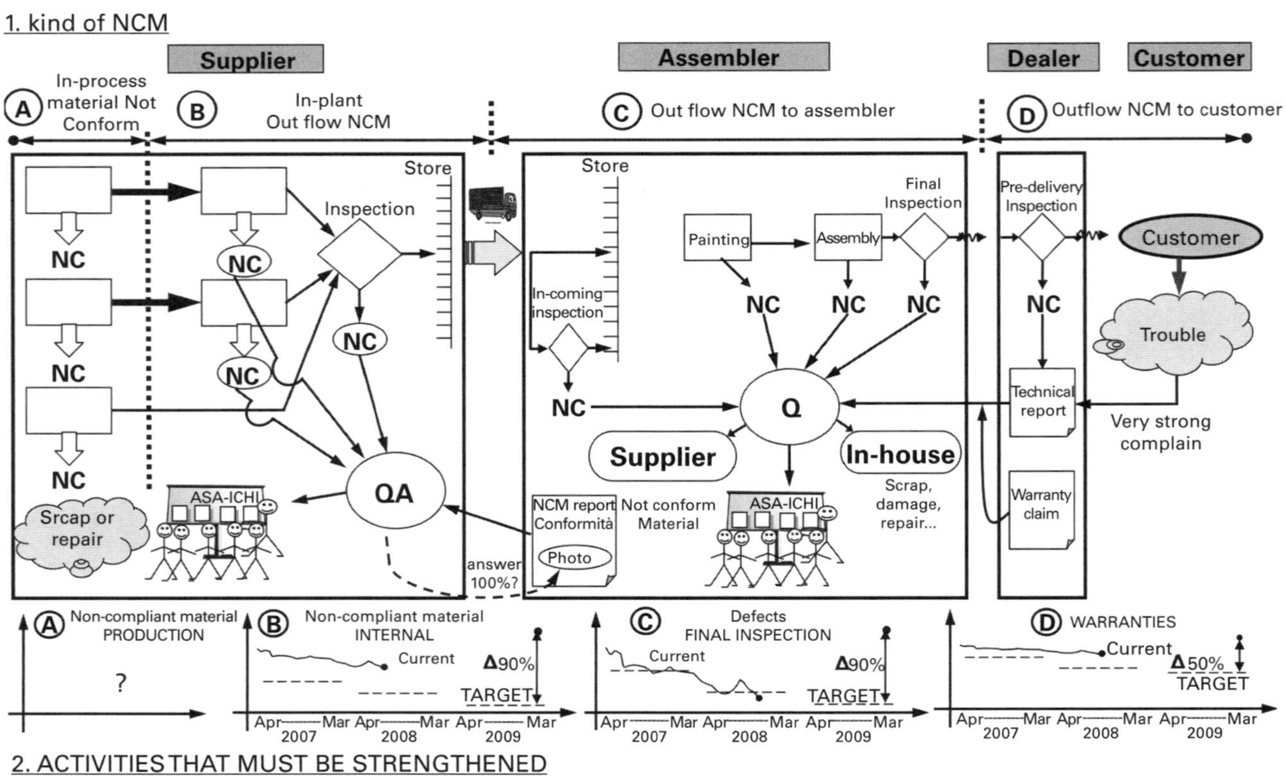

Figure 11.8 NCR performance

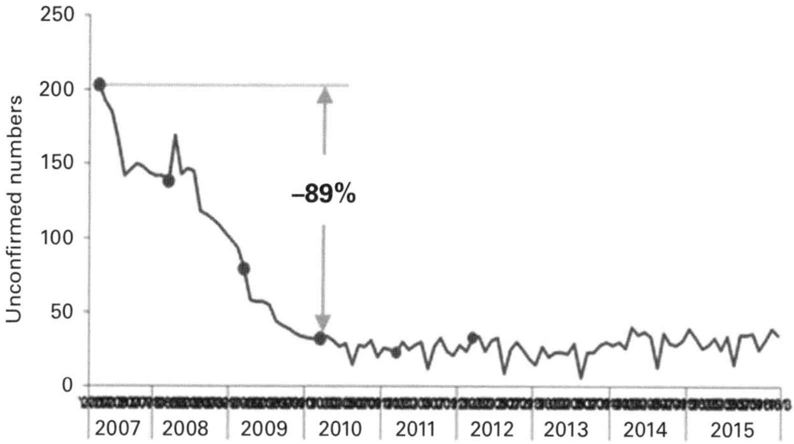

The introduction of *Asaichi* in the main production sites of the suppliers, after appropriate training and implementation processes, allowed them to achieve notable results, not dissimilar to those recorded by TMHMI (Figure 11.8).

Reduction of WCPs (warranty claim paid)

The third area of intervention that was acted upon for the *Dantotsu* concerns the cost of warranties.

This is the area which, aside from involving substantial costs, has a significant impact on the brand image; these are issues that are by nature hard to spot during the inspection and testing stages during production, and they are often related to engineering or supply.

Despite the utmost attention given to development and testing of new products, some quality issues can still arise when the product reaches the market, and are reported as warranty costs. In such cases, it is crucial for the company to be able to identify and act on them immediately with the proper countermeasure, not only to supply the customer with an adequate answer, but also to minimize potential growing costs and to avoid damages to the product image. The target of hitting a 75 per cent reduction of warranty costs in three years looked extremely difficult, if not impossible. The first step was to assign a department in charge of corrective actions related to the warranties being issued, which could be production, R&D or supplies. A dedicated area, *Obeya* (Figure 11.9),[4] was set up to give maximum visibility

Figure 11.9 A dedicated *Obeya* for the management of warranties

to the activities and issues reported by our customers, allowing all the available information to be visualized (Toyota's visual management). A weekly meeting attended by top management was then organized to assign and discuss the main problems and find the appropriate countermeasures.

Obeya's success was such that a new one was created for the development of new models, where all the information related to the new product is displayed. The project manager organizes weekly project advancement meetings directly inside the *Obeya*.

Pareto analysis[5] was adopted to give priorities and importance to all warranty claims and identify the main intervention areas. For instance, the issues were categorized as:

- *hydraulic* (oil leaks, contamination issues, power steering or distributor faults, etc);
- *electric and electronic* (software issues, control unit faults with error messages, etc);
- *mechanical* (parts breaking under stress, assembly mistakes, failures).

We started to work on the three main issues for each area, taking into consideration the financial impact and their frequency. The materials we used for the analysis and the solutions include blueprints, pictures retrieved from the net and results of dedicated analysis.

Case history

The warranty claims that are harder to identify are related to oil leaks from the fittings of the hydraulic pipes. Usually, the only action performed by the service department was to tighten the leaky fittings, which meant that we had no parts we could examine to identify the cause of the fault. In order to

Figure 11.10 Reports of oil leaks received from the sales network

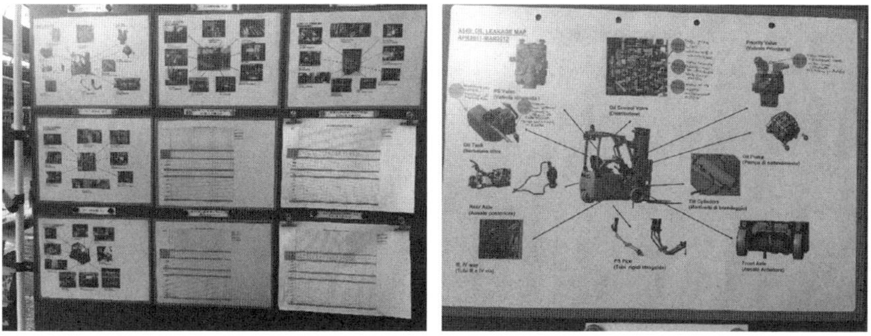

gather further information that could help us establish the issue more clearly, we asked service to indicate the exact location of the leak on a drawing of the part (Figure 11.10). With the data we received, we created a map where each leaking area was marked with a red sticker that showed:

- date of servicing;
- hours in operation;
- sequential number of the problem arisen from the same area.

The adoption of this method quickly allowed us to have a visual depiction of the most affected areas; we adopted electronic screw guns to improve the operation and required the operators to mark the screws that were tightened, so that we could confirm that the operation was properly carried out.

Another method we use to display the issues we encounter, and to assess the performance of the countermeasures that are taken, is 'weak point management' (Figure 11.11). A graph is created for each issue, where the abscissa is the reference month of production and the ordinate the hours in operation. Each notification is manually marked on the graph, indicating the date of entry: this allows us to assess at a glance the effectiveness of the countermeasures that we applied, because a successful solution means that no new claims are reported after the date of implementation.

We needed to have a well-defined vision of our warranty costs, and we wanted to send a clear message to sales networks, distributors and branches: 'quality has no compromises'. For this reason, we requested the return to the factory of each and every faulty part, regardless of its actual worth or shipping costs.

Figure 11.11 Example of weak point management

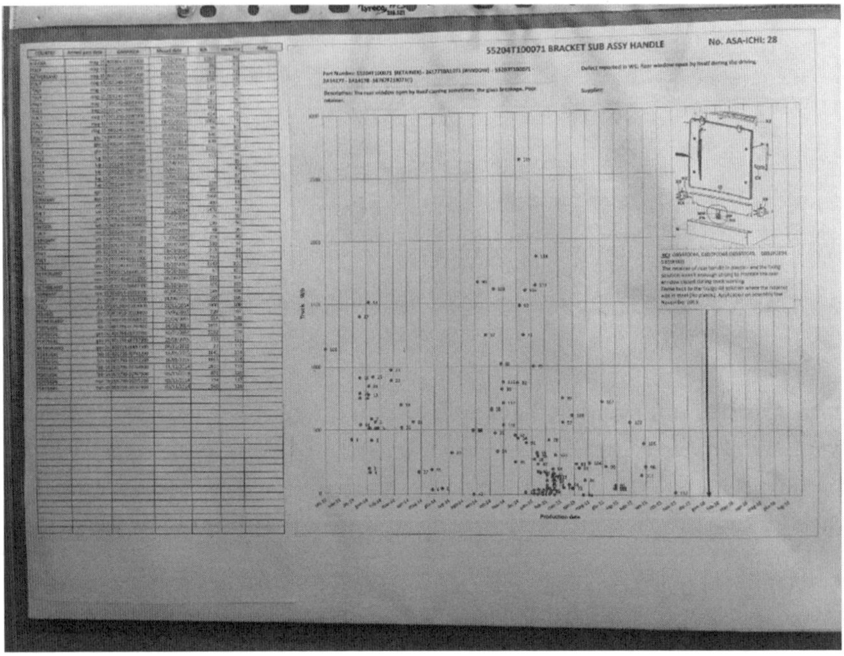

This also allowed us to inspect faults and breakages of low-cost parts that were causing issues to the operations. Not only did we ask for the part to be shipped back to us, but we also asked all service branches to send us as much information as possible about the fault, including pictures and drawings. This was essential to identify issues that were not strictly related to a faulty part. For instance, some of the electronic components that were deemed faulty had no inherent issues, but they malfunctioned when interacting with other parts of the forklift. Finding the real reason entailed a substantial diagnostic effort, performed with the support of the sales and service networks.

With the aim of giving the appropriate importance to warranty issues, the *Obeya* was moved directly to the area where the faulty parts are received (Figure 11.12). In this area, each team member can verify the fault on the part itself and gather their conclusions. If necessary, the suppliers are invited to the meetings to share our thoughts and suggest remedial actions, because they are responsible for a significant part of the warranty costs.

Figure 11.12 *Obeya* of warranties, with the parts that require inspection

A crucial element that can improve the quality of the products is the planning process; the performance of the product is evaluated from the very beginning with prototypes and finite element method (FEM) simulations.[6] The development of a new product must be based on designers' experience, test severity and previous mistakes on older models. All data flow into the process of product development, which is itself structured into eight steps, called design reviews (DRs). Each DR has an approval stage, where management assesses the results of the development of that specific step and approves (or halts) the project. In addition to DRs we published an operation manual that outlines all the steps of product preparation: each activity of every function that needs to be performed during product development is included here. Each duty is assigned to the functions that are responsible for it, while each assignment has its own timeframe, from design to implementation. All the activities and the materials regarding the project are gathered in a dedicated *Obeya*, where meetings are held to discuss the advancement stage of the project. In addition to DR and product preparation, all development stages are followed in close communication with the industrialization, production and supply areas. The first prototypes are assembled jointly by the prototype department and by the production workers, so that any improvement can be identified from the very first development stages. Aside from contributing to improve the products considerably, this solution has drastically reduced the conflicts that are typically found between R&D and operations.

Figure 11.13 Level of warranty costs as 'total costs paid'

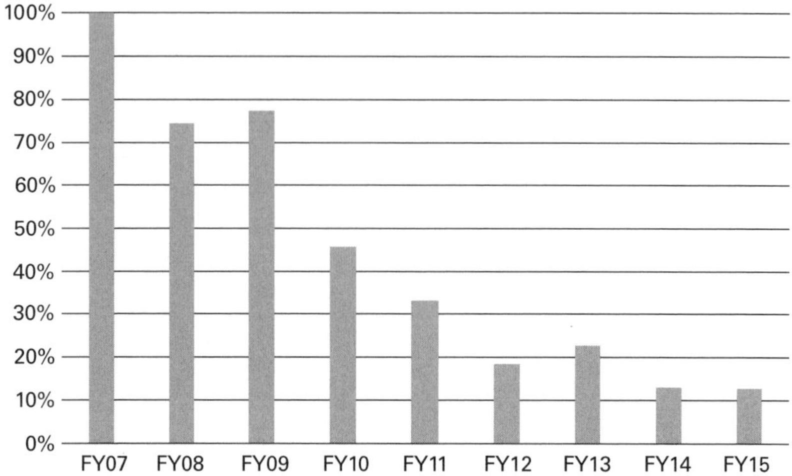

This set of activities succeeded only by involving the whole company, making everyone aware of quality issues, explaining them clearly and giving them the right relevance; this collaboration ensures that with the right analysis tools we are able to implement the most effective countermeasures.

Warranty costs started to decline, despite the volume increase and the addition of new models, hitting –55 per cent in just three years and subsequently reaching an impressive 80 per cent reduction (Figure 11.13).

Notes

1 Problem-solving (which will be described in detail later) is an approach to the solution of a problem that includes various methods.

2 The checkpoint is intended as a station along the production process, where testing and verification are performed.

3 Once I decided to give the whole department the responsibility for the solution of a problem, but Toyota's senior advisor, who was my referral for the project, argued that the department would have never acted, because a department is made by everybody and by nobody. If I had picked a specific person, I could be able to identify who is responsible for the solution.

4 *Obeya*, Japanese for 'big room'. The term derives from the Prius project, which was assigned to a young chief engineer, Takeshi Uchiyamadarande, who relied on a transversal team. In order to facilitate communication and shorten times,

the team was placed inside the *Obeya*, and all the material regarding the development process was hung on the walls. This method was so successful that it was applied to all subsequent processes and extended to the entire group.

5 Pareto analysis consists of a technique that allows the identification of the main causes responsible for most of the issues. It is also called 80/20, since Pareto calculated that usually 20 per cent of causes generate 80 per cent of the problems.

6 FEM is a mathematical method used to find approximate solutions to problems, described in differential equations, to the partial derivative by reducing the latter to a system of algebraic equations. It is widely used in engineering, often to simulate the stress state of a part with the aid of dedicated software, pointing out the areas where breakages might occur. TMHMI, in addition to the use of FEM, undertakes specific tests with working prototypes, inside which are inserted extensometers that measure the deformation level of the parts under stress.

12 Visual management

During a visit to our plant by Toyota's top management, I was about to explain how the production activities were structured when my guest stopped me: 'A factory is a Toyota factory if a visitor can understand it without anyone having to explain it.'

This quote encloses many concepts that we have previously treated: the flow of materials, clearly visible production lines, specific areas allocated for each activity, and most of all the visual management approach, which in Japanese is called *Mieruka*.[1]

A fundamental concept of TPS is that all the activities, the KPIs and the projects cannot be just a mere working tool for management, but they must be visible to anyone. Its potential can be exploited not only by using all the tools developed along the years, but also with the involvement of all the company employees, in particular those who are directly involved in production activities. Displaying clearly to everyone all KPIs, projects and activities brings two important benefits: direct involvement of all the personnel working on operations, and the presence inside the factory of all management functions.

The application of visual management has been one of the most difficult tasks we had to face, because it can be effective only if it becomes an essential working tool for everyone. We had to reach our goal, and avoid relegating the visual management approach to be just a fad that needs to be implemented during the visits of Toyota representatives. It took great determination and perseverance from myself and my management team to achieve the desired result.

Let's begin by describing an operative visual management; those who work directly on the production lines are in charge of its management, while the GL is responsible for its operations and updates (Figure 12.1).

Figure 12.1 Example of a visual management board run by a GL

This area is usually placed next to the production lines, and it contains all information, KPIs and graphs that support the GL in the daily management of their work. As previously described, the training of group leaders is coordinated directly by Toyota's headquarters in Japan; this means that the layout and information on the boards are structured in the same way in all the plants of the group.

One section is dedicated to 2S, the cleaning and standardization activities of the stations. An evaluation system is used to rate all the stations, indicating the areas in need of improvement (a radar chart shows the progress status based on a set target); the drawing of the layout of the department concerned can be used to signal, using a red sticker, the areas in need of intervention (Figure 12.2). All the related activities are displayed with a Gantt chart that includes timeframe and progress stage, also indicating the name of the person in charge of the activity.

One section is dedicated to the skill matrix. In the previous chapters we stressed the importance of the operators' training, and the crucial role it plays in quality and safety. Knowing the 'education' and autonomy levels of each operator is fundamental for a proper management of the area. A simple

Figure 12.2 2S activities on the production lines

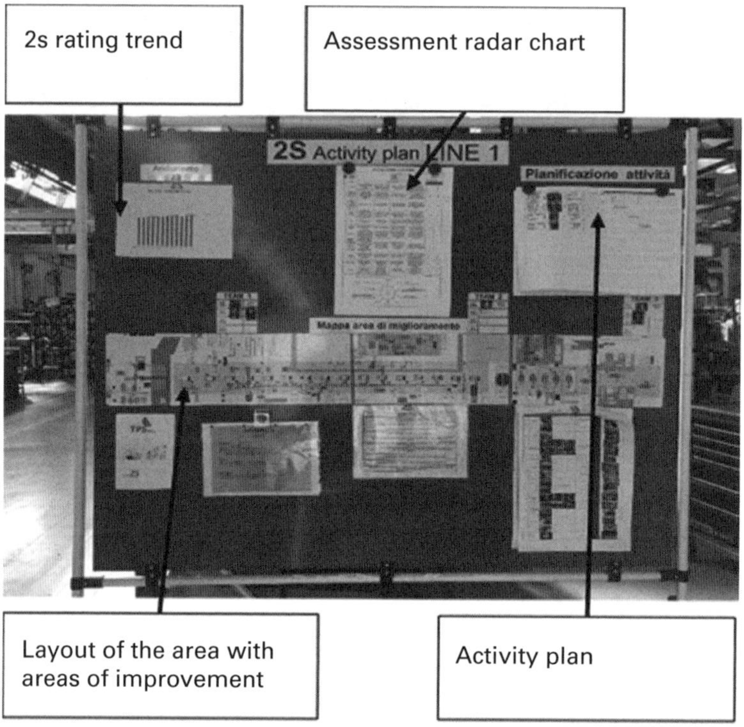

matrix includes the names of the operators and their level of training and knowledge of one or more workstations. This simple tool allows us to:

- simplify the allocation of workstations in case of vacation/time off/sickness of operators;
- see clearly who are the more experienced and knowledgeable workers;
- set up a training plan that can promote and implement multi-functionality.

In addition, coloured magnets help to quickly identify the knowledge and autonomy level of each worker for a specific workstation: from red, used for an operator under training, to blue, which indicates a great level of experience of the operators, up to the point that they can be considered as trainers themselves (Figure 12.3).

Another GL-run visual management area is the 'change point management' (Figure 12.4). This tool was developed to highlight any changes, usually related to external or organizational causes, that can affect quality, safety and productivity.

In this area the organization is depicted using pictures of the operators, including those of GLs and TLs, each with their own assigned areas; for

Figure 12.3 Example of a skill matrix of the operators on Line 1

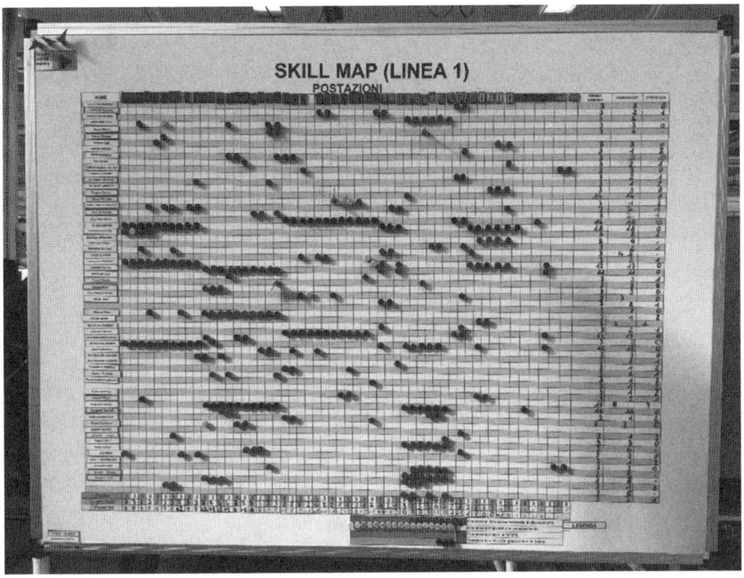

Figure 12.4 Example of a change point management area

instance, in the change point management illustrated in Figure 12.4 there are three TLs in charge of a section of line split into Team 1, Team 2 and Team 3. The productive area layout of each GL and TL shows the pictures of the operators that are assigned to each station for that day, so that it is

Figure 12.5 Layout of production lines, workstations and operators

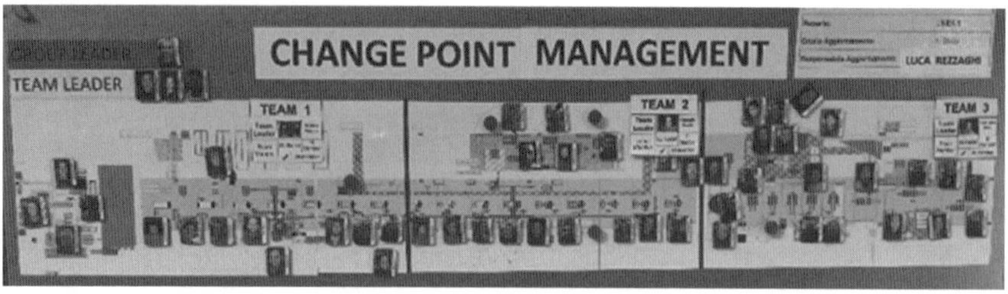

Figure 12.6 Management of workers' attendance

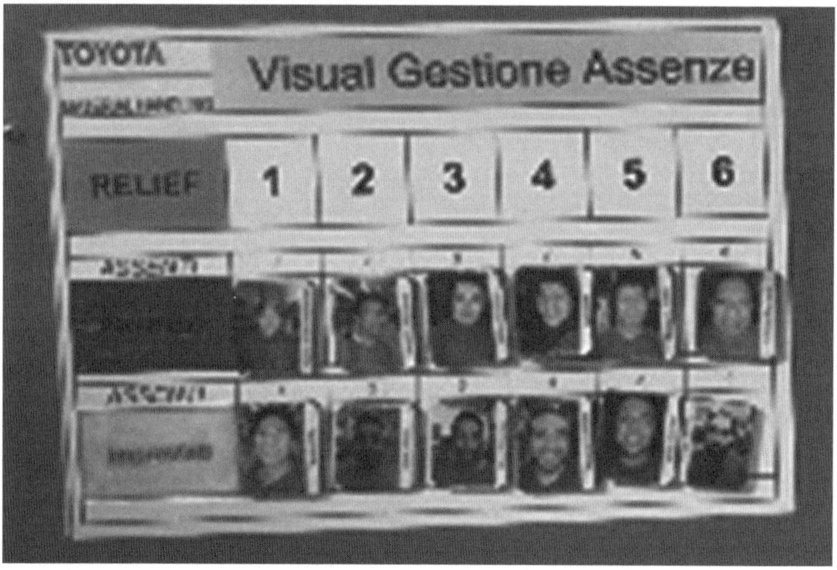

possible to verify, with the support of the skills matrix, whether the assignment is consistent with the competence level of the operator (Figure 12.5). Finally, the pictures of the operators who are not at work are placed in a separate area, where they are classified as 'planned absence' (due to work permits of any kind) or 'unplanned absence' (due to sickness or injury). It also includes the pictures of the workers who replaced those who are not on site (Figure 12.6).

The CPM of each station, in addition to displaying the above information, highlights the changes that might cause situations of potential risk. The changes are classified according to 4M:

- *Machine*: a **blue** magnet is placed on the area of the workstation affected by the changes (such as screw guns or new equipment).

Figure 12.7 Example of 4M management

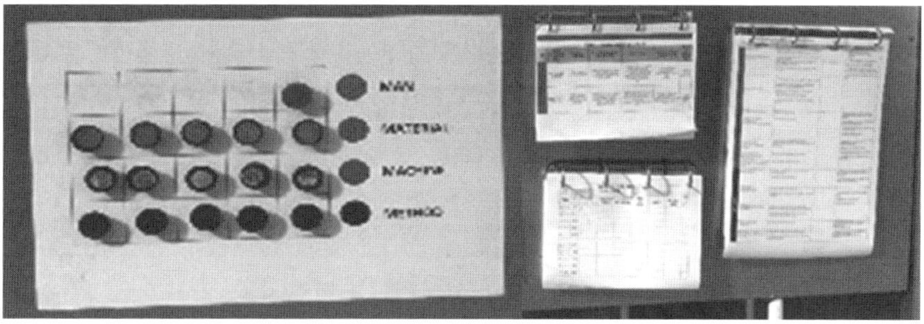

- *Material*: a **green** magnet is placed on the area of the workstation where materials have been changed (ie if there is a new part in the assembly list of the station, or if the layout of the feeding of materials has changed).
- *Method*: a **black** magnet is placed on the area of the affected station to indicate variations of the working method, for example a new assembly sequence.
- *Man*: a **red** magnet is placed on the area of the affected station when the potential risks are caused by the operator, for instance if they are being trained.

Each of the 4M has a colour-coded table where the descriptions of the causes for each potential risk are indicated (Figure 12.7).

A specific area is dedicated to the main KPIs, showing the performance for the following elements:

- Q (*Quality*): the number of fault-free forklifts that are delivered to the next process, on a daily and monthly basis.
- C (*Cost*): compliance with the daily production target, hours spent for each truck produced, any downtime of the lines for whatever reason.
- D (*Delivery*): production volumes reached on that day and on previous months, compared with the target (Figure 12.8).

Furthermore, the progressive target volumes are assessed every two hours, taking note of the number of assembled trucks and comparing the figures with the target. Each production problem has its own cause displayed on a graph, so that it is possible to identify immediately the main issues that occurred. Downtimes due to technical problems are not only recorded, but also assessed in a dedicated area, with the countermeasures that have been established (for the short term, to restart production, and for the long term, so that the problem no longer arises).

Figure 12.8 Example of a daily delivery graph

DELIVERY (D): UP TIME RATIO - DAILY

LINEA 1
Aprile 2016
PIC: Rezaghi Luca

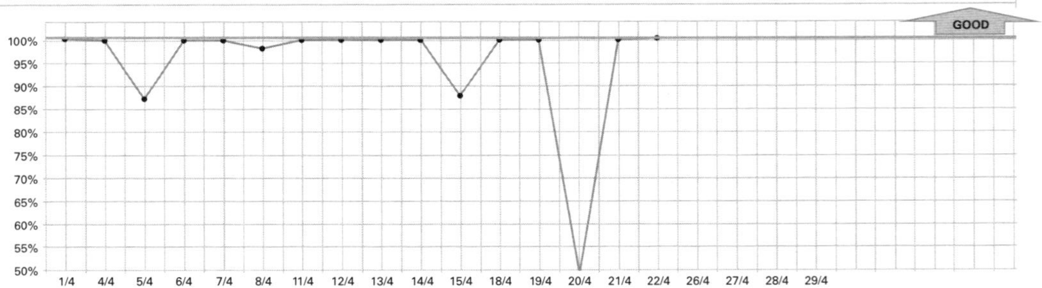

			Data	1/4	4/4	5/4	6/4	7/4	8/4	11/4	12/4	13/4	14/4	15/4	18/4	19/4	20/4	21/4	22/4	#REF!	26/4	27/4	28/4	29/4	0/1	0/1	0/1	Totale
UNIT/DAY	A	Target	Obiettivo	48	55	55	55	55	47	55	55	55	55	48	55	55	55	55	48	#REF!	55	55	55	48	0	0	55	**#REF!**
	B	Actual	Totale giorno	48	55	48	55	55	47	55	55	55	55	42	55	55	27	55	48	#REF!	0	0	0	0	0	0	0	**810**
	C	Prod. Loss	Persi giorno	0	0	0	0	0	0	0	0	0	0	0	0	0	0	0	0	0	0	0	0	0	0	0	0	**41**
WORK TIME	D	Normal	[min]	400	460	460	460	460	400	460	460	460	460	400	460	460	460	460	400	#REF!	0	0	0	0	0	0	460	7120
	E	Actual	(D-I) [min]	400	460	400	460	460	400	460	460	460	460	460	460	460	220	460	400	#REF!	0	0	0	0	0	0	460	6880
TAKT TIME	F	Takt Time	[min]	8,36	8,36	8,36	8,36	8,36	8,36	8,36	8,36	8,36	8,36	8,36	8,36	8,36	8,36	8,36	8,36	8,36	8,36	8,36	8,36	8,36	8,36	8,36	8,36	**8,36**
UP TIME	G	Up Time	F x B [min]	400	460	460	460	460	460	460	460	460	460	400	460	460	460	460	400	#REF!	0	0	0	0	0	0	460	0
	H	Up time Ratio	(G/E) [%]	100%	100%	87%	100%	100%	98%	100%	100%	100%	100%	88%	100%	100%	49%	100%	100%	#REF!	0%	0%	0%	0%	0%	0%	0%	0%
DOWN TIME	I	Total down time	(L ~ M) [min]	0	0	60	0	0	0	0	0	0	0	60	0	0	240	0	0	#REF!	0	0	0	0	0	0	0	**#REF!**
	L	Union assembly	Assemblea	0	0	60	0	0	0	0	0	0	0	60	0	0	0	0	0	#REF!	0	0	0	0	0	0	0	**#REF!**
	M	Strike	Sciopero	0	0	0	0	0	0	0	0	0	0	0	0	0	240	0	0	#REF!	0	0	0	0	0	0	0	**#REF!**
	N	Quality check	Controlli qualità	0	0	0	0	0	0	0	0	0	0	0	0	0	0	0	0	#REF!	0	0	0	0	0	0	0	**#REF!**
	O	Line down time	Fermo Linea	0	0	0	0	0	0	0	0	0	0	0	0	0	0	0	0	#REF!	0	0	0	0	0	0	0	**#REF!**
	P	Jig trouble	Problemi Attrezzature	0	0	0	0	0	0	0	0	0	0	0	0	0	0	0	0	#REF!	0	0	0	0	0	0	0	**#REF!**
	Q	Quality defect	Difetti qualità	0	0	0	0	0	0	0	0	0	0	0	0	0	0	0	0	#REF!	0	0	0	0	0	0	0	**#REF!**
	R	Parts shortage	Mancanza materiale	0	0	0	0	0	0	0	0	0	0	0	0	0	0	0	0	#REF!	0	0	0	0	0	0	0	**#REF!**
	S	Manufacturing	Problemi al montaggio	0	0	0	0	0	0	0	0	0	0	0	0	0	0	0	0	#REF!	0	0	0	0	0	0	0	**#REF!**

Among all the information shared and displayed inside the plant, a very important section is dedicated to safety. Each GL-managed area has its own indicator, summarizing its safety levels. There are two kinds of potential events: *injuries*, which involve and damage a human resource, and *near-misses*, episodes that did not cause injuries but that could have if the dynamics had been different. In both cases a report is drawn, illustrating the situation and immediately implementing the countermeasure. The safety department is next to act, gathering all the available information to establish the causes of the event. In collaboration with operators, TLs and GLs, it sets a possible countermeasure that must be immediately implemented. If this is not possible, perhaps because it requires the purchase or the modification of a piece of equipment, a temporary measure is adopted while the conditions are applied. Any injury is identified and displayed with a red dot placed in the area where it occurred; the same logic applies to the yellow dots that indicate a near-miss. With the aim of sharing all 'bad experiences' inside the Toyota group of companies, injury-related reports are shared among all the production sites. The use of a visual scheme such as the one shown in Figure 12.9 helps to determine which area or station has been exposed to the highest number of risks. The display is integrated with a chart showing the number of injury-free days and their trend, with a calendar outlining each accident-free day.

The analysis and the potential improvements that can be applied in the cycle of a production workstation are identified with the adoption of a *Yamazumi* chart[2] (Figure 12.10). This tool, created to maximize efficiency and minimize waste, can display the time required to perform every operation of each workstation. The operations are classified as:

- *Added-value operation:* required operation, without which the unit would not become a finished product.

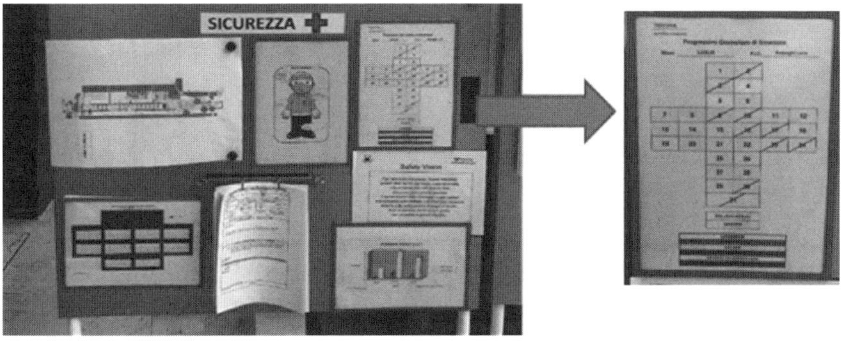

Figure 12.9 Example of visual indicators dedicated to safety

Figure 12.10 Example of a Yamazumi chart

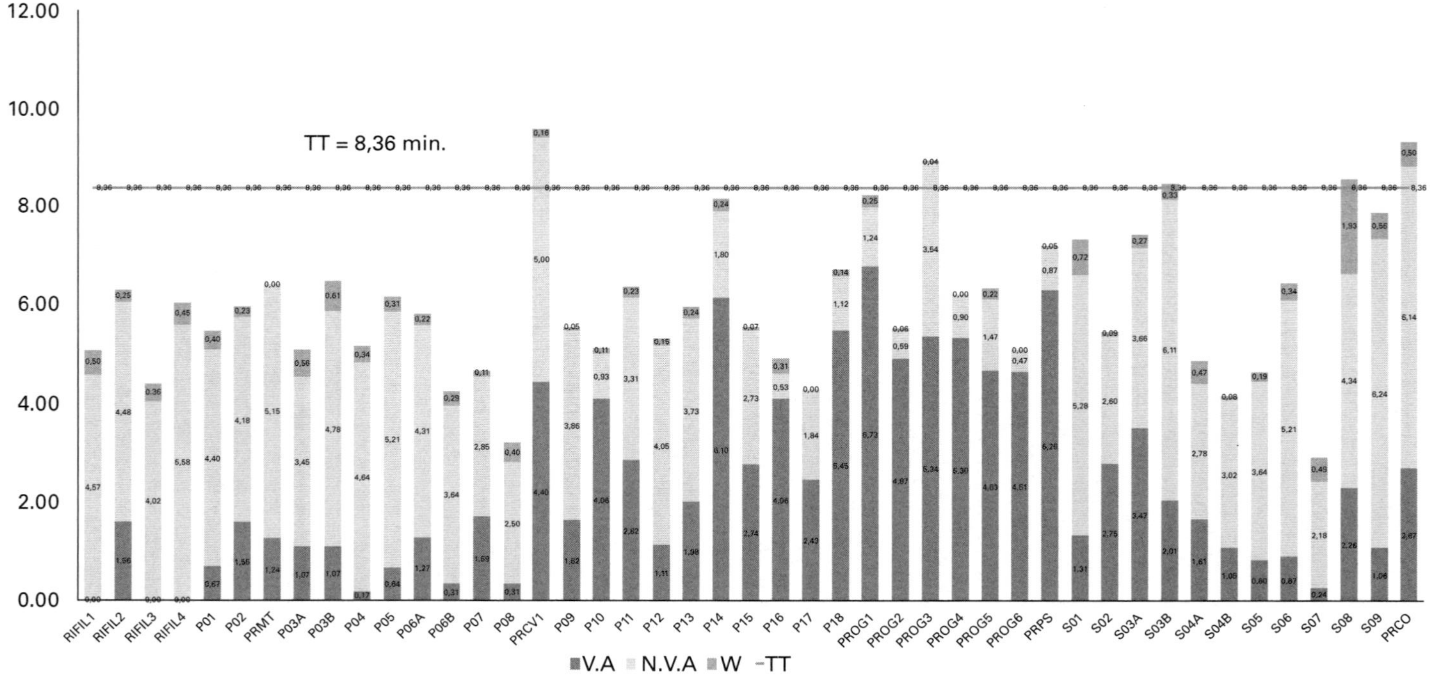

- *Waste-reducible operation:* it is a required operation but it must be minimized, since it does not add value for the customer.
- *Non-added-value operation (Muda):* an operation that constitutes a waste (such as operator motions, repairs, etc).

Once times are recorded and classified with the above criteria, they are displayed on a histogram which shows the total time needed to complete the cycle at every station.

The histogram also reports, for reference, the takt time assigned to the production line.

The criteria used to classify the operations with added or non-added value are crucial to identify potential *Kaizen* that could improve the station's efficiency. Figure 12.11 shows some examples of the Toyota classification method.

Some of the classifications are proof of the great efforts made to identify any potential improvements. For example, the action of passing the wiring and cables through the frame is considered as a reducible waste. At a first glance it looks like a required activity, because the forklift would not work without it. However, its classification highlights that the system used to lay cables and wiring can be improved. Having a fixed layout for laying all the cables, with preset anchoring points, makes the operation substantially faster than using cable ties in points that might differ in each truck. Its classification as a 'reducible' *Muda* indicates the need to assess the procedure by the work team, in order to find ways to optimize the activity. To determine and visualize the operator movements, we implemented a dedicated recording system called the standard motion sheet (Figure 12.12): it shows the station layout with the materials and the movements made by the operator. Each operation is classified as an activity performed in a fixed or a moving position, also showing the time needed to carry it out.

The first applications of this method revealed that on most of the stations, 30 per cent of the operator motions were needed to retrieve the parts. It was immediately clear that if we wanted to minimize unnecessary motions, we needed to relocate parts and equipment, placing them in line with the production cycle. We created a *Yamazumi* chart for the purpose of efficiently displaying the operator movements, using colours to pinpoint the operator workstation related to the assembly stage of the forklift. This way, if we look at the tasks to be completed while the operator is sitting on the driver seat, the assembly cycle of the front side is indicated in green, in blue for the left side, yellow for the right and grey for the back (Figure 12.13).

Figure 12.11 Example of activities classification for the Yamazumi

CATEGORY	N°	DESCRIPTION	DETAILS and EXAMPLES
VALUE ADDING OPERATIONS	1	Parts assembly	
	2	Screws tightening	
	3	Connections	
	4	Tacking/Welding	
	5	Painting/Spray	
NON-VALUE ADDING OPERATIONS	1	idle movements/waiting times	Picking materials, tools...
	2	Checking files or lists	Reading, parts list with several reference codes...
	3	Turning around	
	4	Bringing materials back	
	5	Moving tools or equipment from one hand to another	
	6	Moving to pick parts and tools	
	7	Moving while holding parts and tools	
OPERATIONS WITH REDUCIBLE WASTE (that can be improved)	1	Preparing the parts	
	2	Starting something	Computers, signals...
	3	Installing wires and pipes	(ex. applying cable ties)
	4	Adjustments (ex. If the screw does not fit in the thread)	
	5	Applying and removing stickers, labels...	
	6	Reading lists or files	
	7	Returning boxes and pallets	
	8	Kanban collection	
	9	Supply of screws of different types	

Reporting by colours the cycles and times on the graph allows the sequence of operator movements caused by the lack of optimization of the assembly cycle to be displayed. Ideally, grouping the activities by forklift areas allows the movements that the operator has to make during the assembly to be reduced.

The *Yamazumi* chart was originally implemented by the times and methods department, but it showed its real potential when its management was passed over to the TLs of the production lines. This cultural shift gave the opportunity to those who work daily on direct activities to learn how their stations are balanced; as a consequence, they acquired the knowledge they needed to suggest and implement the *Kaizen* actions to improve efficiency. The *Yamazumi* chart can be enhanced with the use of magnets, showing on

Figure 12.12 Example of a standard motion sheet

Standard Motion Sheet	Job Content	From prendere chiavi / To ritorno a carrello	Page # 1 1	Model blitz	Process # 6	Date 27/02/2008	Revision / /

T.T	17,14	min.
C.T	13,78	min

Seq #	Operation name	Time M	Time W	Seq #	Operation name	Time M	Time W
1	prendere chiavi	13,7		41	togliere tappi tubo		11
2	prendere viti	9,51		42	ritorno a telaio		4.4
3	andare al telaio	5,51	9,3	43	montare tubo	19	
4	impuntare vite	19		44	serrare	34	
5	leggere scheda	12,4		45	togliere tappo tubo		7.4
6	prendere l'alimentatore	9,54	7,14	46	ritorno a telaio		4.9
7	ritorno a telaio		6,7	47	montare tubo	14	
8	gettare imballo	7,48	3,61	48	serrare	12	
9	ritorno a telaio		7,15	49	riporre chiave		3.5
10	montare scheda I/O	66		50	prendere pennarello	5.4	3.6
11	prendere avvitatore		7,07	51	certificare serraggio	5.4	
12	ritorno a telaio		5,75	52	prendere fascette		4.5
13	serrare	24,5		53	sistemare cablaggio	28	
14	riporre avvitatore		5,74	54	tagliare	7.7	
15	ritorno a telaio		5,7	56	gettare	4.5	
16	inserire cablaggio	57		56	ritorno a telaio		4.5
17	collegare I/O	14,5		57	montare fascette	20	
18	collegare l'alimentatore	12,7		58	tagliare	7.9	
19	montare alimentatore	92,3		59	gettare	7.3	
20	prendere il distributore	6,3	6,12	60	ritorno a carrello		11
21	posizionare il distributore	15,5		61			
22	impuntare distributore	21		62			
23	riporre il braccio		6,77	63			
24	prendere avvitatore		4,4	64			
25	ritorno a telaio		5,26	65			
26	serrare	11,7		66			
27	andare a riporre avvitatore		6,24	67			
28	prendere tubo D,E,F,L	10,2	5,86	68			
29	ritorno a telaio		7,7	69			
30	togliere i tappi	10,1		70			
31	ritorno a telaio		5,1	71			
32	posizionare un tubo	18		72			
33	prendere grasso	4,78		73			
34	applicare	3,56		74			
35	riporre grasso	5,2		75			
36	montare tubi	27,2		76			
37	prendere avvitatore		3,86	77			
38	serrare	14,5		78			
39	riporre avvitatore		3,83	79			
40	ritorno a telaio		3,15	80			
					Sub-total	656	170
					Total	827	

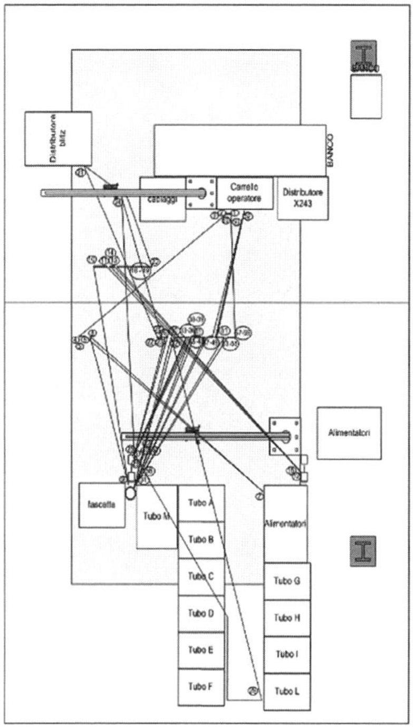

Figure 12.13 Identification of the working areas

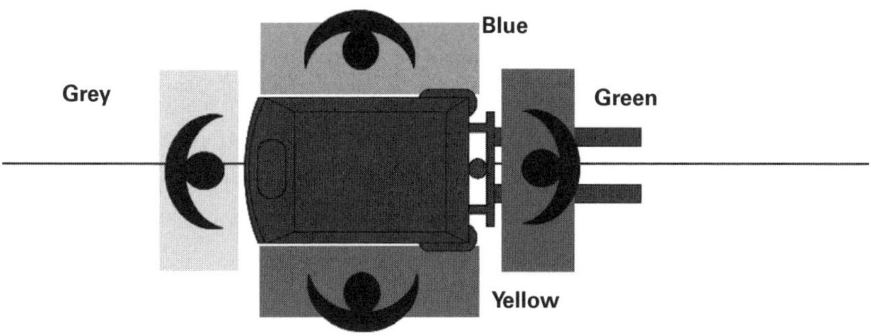

a timescale the time it takes to execute an operation. Magnets allow, with a quick glance, the transfer of activities between stations to be assessed in order to improve their balance, to inspect the operator movements and to spot any workstations in over-takt due to specific product configurations. Visual management is not limited to the management of operations during production,

Figure 12.14 Example of visual management applied to projects

but has been extended to all the ongoing projects of the company (Figure 12.14), including those related to the development of new products. In the operations area, each project displays all the information required following a strict standard, its board placed directly in the reference area.

At the beginning of the procedure, in order to prepare the A3 report,[3] we collect all the data we need to define the current state of things and the gap that we need to fill in order to achieve the desired results. This is followed by a very detailed analysis that focuses on several factors; finally, the results of the analysis and the information supplied by the operators help identify a potential solution. It is important to begin with the history and the reasons behind a new project.

If it is necessary to make modifications in the layout of the production areas, we start by highlighting all the critical spots, which we identify and mark on a drawing of the areas. Usually, this is done with coloured stickers placed on the identified spots (Figure 12.15), or with Post-Its that briefly outline the problem. Countermeasures are studied for each of the critical spots; the suggested actions start the development of a new solution that, if appropriate and applicable, is integrated in the new possible layout project. Each project has its own KPIs, indicating the starting conditions, the target

Figure 12.15 Example of visual management, indicating the critical points of the current solution

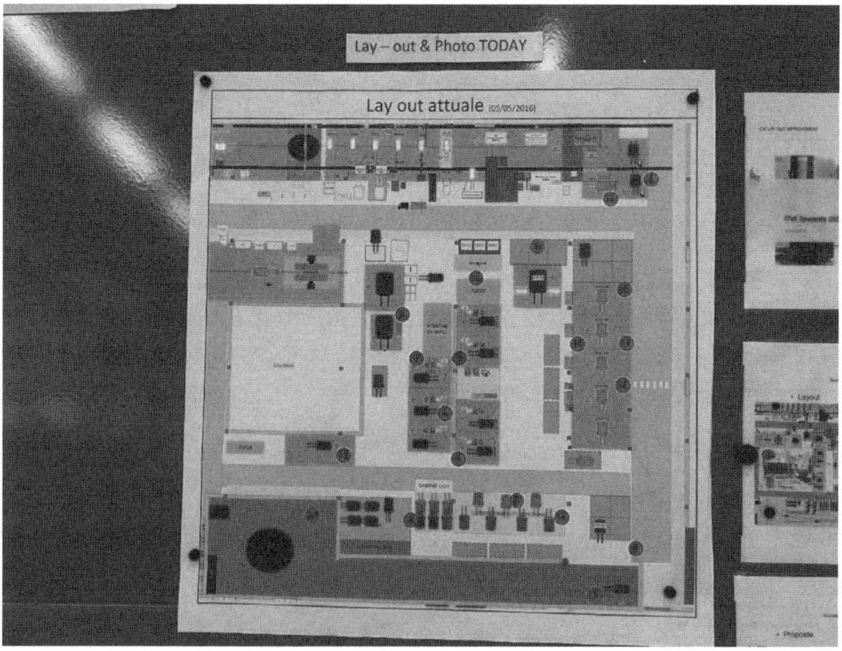

and the progress towards the expected results. Targets are classified under QCDESM categories, which will be the subject of separate chapters.

Such complete visibility of the progress stage of the project, directly in the reference area, apart from involving the workers in the improvement effort, is the ideal method to inform and make the operators aware of the modifications that must be applied by the company.

The following items are always indicated on all the projects:

- work team;
- target and KPI;
- current situation;
- list of gaps between current situation and target;
- several hypotheses of future solutions, such as different layouts;
- investments and payback analysis;
- weighted assessments of the proposed solutions;
- chosen solution;
- work plan.

Notes

1 *Mieruka* (literally 'manage with the eyes, make visible') is one of the main points of TPS philosophy. With *Mieruka* are displayed the main KPIs, current projects and corrective actions.

2 The *Yamazumi* chart is a bar chart where the working cycles are split into added-value and non-added-value activities.

3 A3 report is a method developed by Toyota for projects and problem-solving. Its name derives from the fact that all the information is displayed on an A3 sheet, where each area covers a specific topic. I really appreciated its use, because it's a perfect template for analysis and representation.

13
Kaizen and efficiency

TPS was created with the main goal of reducing and eliminating waste (*Muda* in Japanese). A constant and thorough effort made possible the development of the tools that this cultural approach implemented, making it famous worldwide. There are seven different kinds of *Muda* in Toyota, but overproduction is seen as 'the father of all *Muda*', because it is a direct cause of all other kinds of waste (Figure 13.1).

Overproduction creates more *Muda* because it raises the amount of stock in inventory, which in turn needs more transportation and operators, more areas for storage, longer pick times and consequently longer waiting times. In the event of quality defects, more material needs to be retrieved from storage, which represents another waste of time; if the materials are stored for too long, they can deteriorate. Finally, overproduction is seen as the worst *Muda* of all, because it hides the real issues that lay behind.

Although the labour costs in Japan are not exactly cheap, more than 5 million Toyota vehicles are produced in the country every year and exported worldwide. Moreover, a sizeable percentage of the parts used by Toyota factories located abroad is made in Japan. This proves how an excellent system of production and logistics, where quality and efficiency are constantly at the core of the employees' efforts, can compensate for high labour costs and offer competitive prices.

One of the key concepts of TPS is defined by the Japanese term *Kaizen*. The word is composed by **Kai** and **Zen**, two ideograms (改善) meaning 'change' and 'better': changing for the better.

Toyota sees *Kaizen* as small, gradual improvements, suggested by people within the company, which usually need a very small investment to be applied. A *Kaizen* from the operators who work directly on production activities are particularly appreciated. Often the *Kaizen* is erroneously associated with strategic, complex projects, while sometimes a technology-first approach can cause unnecessary complications. Investments in equipment, tools or other activities bringing substantial modifications, step changes, are called *Kaikaku* (radical changes). For instance, a new layout with the introduction of new machines and technology is a *Kaikaku*. Once the *Kaikaku* is implemented, its efficiency is maximized with *Kaizen* (Figure 13.2).

I belong to a category, the engineers, where technology is often seen as the tool to solve all possible problems. Engineers can draw complex and expensive solutions to deal with even the most trivial issues. I have often rejected some of those complicated brainstorms because they were not in line with

Figure 13.1 Seven kinds of waste

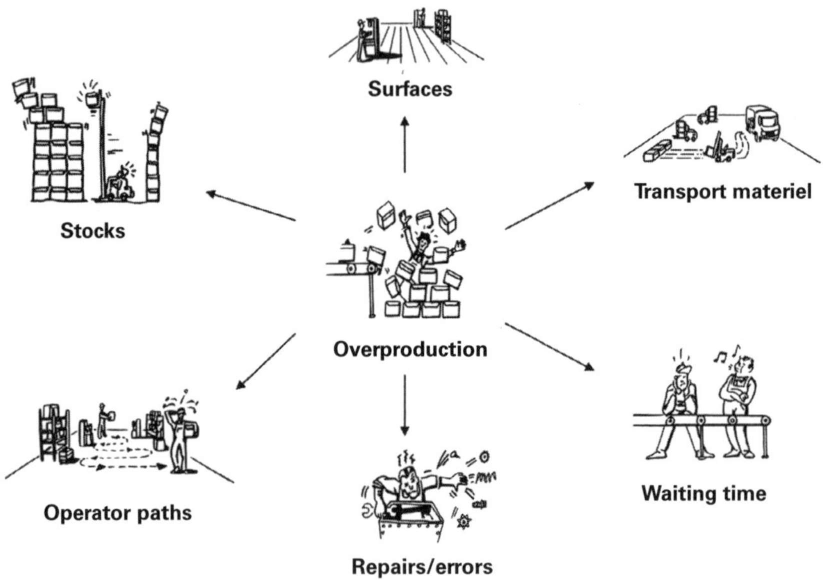

Figure 13.2 *Kaizen* versus *Kaikaku*

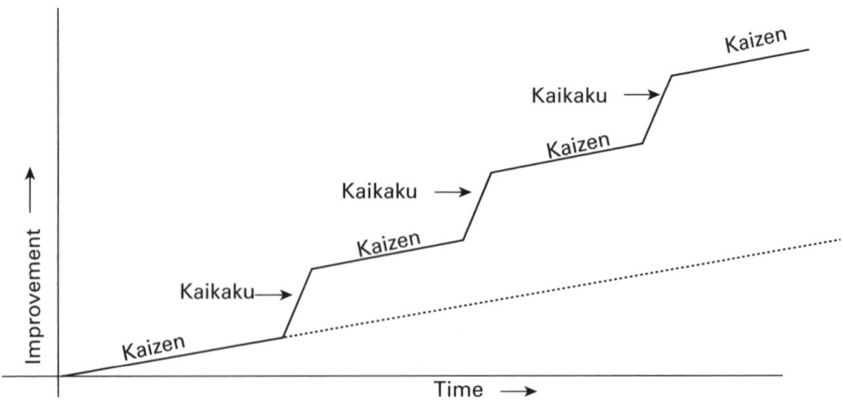

the simplicity principle of *Kaizen*. There were other occasions when we tried repeatedly to find simpler ways, and we discovered some amazingly efficient and straightforward solutions. The guideline that must be followed for any project is 'No People, No Money, No Space'.

I recall one occasion when we had to face a quality problem caused by tightening operations that were missed by the workers. When we analysed the data of the defects, we realized that a potential quality risk was the interruption of production activities. For example, if the operator could not finish performing the activities at the workstation, because the shift was over or for their lunch break, the risk was that they, once returned to work, would not remember at what stage they left the current activity. The idea, actually, the *Kaizen*, came from a GL, who built a flag with a ribbon and a magnet. Each station received a flag, so that the operator could place it on the last tightening carried out, leaving a clear indication when the activity resumed, even if it involved a different operator (Figure 13.3).

Another important aspect is the involvement of the operators working in production activities. History and social conventions had erected an invisible barrier that divided white- and blue-collar workers. This barrier hinders

Figure 13.3 Example of *Kaizen* to indicate the last tightening performed before interruption

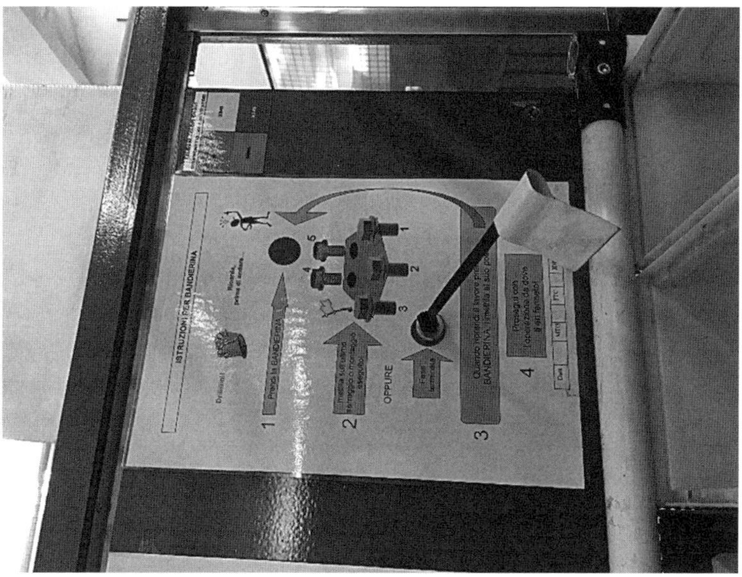

communication between the sides, restricting them to carry out only their own specific tasks: production engineering is the sole responsibility of organization and technology, while operations is limited to the execution of instructions. Furthermore, a barrier creates conflicts between those who decide what to do, and those who actually have to do it.

Kaizen Kobo

The consolidation of *Kaizen* and the increasing involvement by people within the company, who were the first to contribute with their ideas, made it difficult for us to purchase or develop the equipment that could fully support all the *Kaizen* that were suggested. We searched the product ranges of a large number of suppliers, and it was always very hard to find the tool, the equipment, the solution that could fully satisfy our needs. Once again, a trip to Japan proved to be crucial. They already faced and solved the problem, creating an area for the production of small equipment: the *Kaizen Kobo*.

Compared with a classic tooling area, the parts used in the *Kaizen Kobo* for the creation of equipment are standard and simple (Figure 13.4). Using

Figure 13.4 *Kaizen Kobo* area

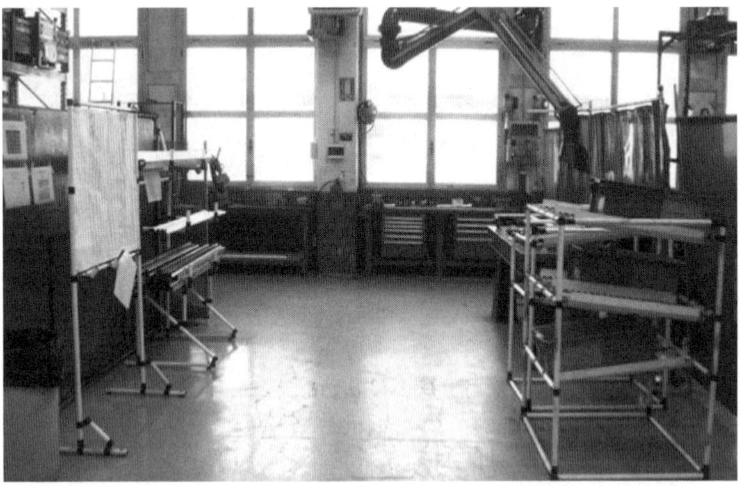

Figure 13.5 Example of a roller conveyor used in a *Kaizen Kobo*

a limited range of wires, tightening systems, wheels and other components, we were able to produce internally the best part of our equipment, with the added bonus of making everything exactly to our requirements. The simplicity of the *Kaizen Kobo* allows the operators to assemble their solutions themselves. Dozens of tools have been quickly created to support the development of *Kaizen* (Figure 13.5).

14
The bottom–up approach to ideas

We have so far illustrated the tools required to apply the *Kaizen*, such as the *Kaizen Kobo*; we have also outlined the core philosophy: simple, effective ideas that mainly come from the operators. The latter factor is the most complex to establish, because it requires a substantial cultural shift. The company has a primary role in this transformation, creating and promoting an environment that stimulates ideas, where people are encouraged to express their opinions and solutions. The concept implies the identification of what the customers perceive as 'value', including the internal customer, the downstream process; this is a fundamental aspect to help identify and reduce waste, improving efficiency and service levels. This task is not only assigned to management: each employee can spot inefficiencies during their daily tasks, often caused by redundancies or non-optimized processes.

The company must train its employees in the identification of waste and of customer value, supplying them with the tools they need to fix the issue themselves. Achieving such a result means an efficient company, with highly motivated employees.

The flow of ideas must reach the managerial levels in order to become effective: their involvement gives the appropriate importance to the activity in question, and ensures the right leverage to support their implementation.

Projects are usually assessed on the basis of the following macro-categories:

- **Q (Quality):** impact on quality.
- **C (Cost):** cost reduction.
- **D (Delivery):** impact on lead time, production capacity and productivity.
- **E (Environment):** impact on the environment.
- **S (Safety):** impact on safety.
- **M (Moral):** improvement of corporate climate.

The last category was recently added and proves how corporate climate is a must for the Toyota group. HR has a key role: it is an integral part of the changing process and is in charge of a smooth and easy transition, promoting training and initiatives aimed at involving the employees in an effort to sustain a positive corporate climate. The main HR indicators used by the group include employee turnover, injuries and absenteeism percentages. Those KPIs are so important that they are shared monthly among all the factories worldwide. There is, however, a crucial and irreplaceable role played by the Japanese top management and their *Genchi Genbutsu*, their periodic visits to the group's companies and production sites: this is by far the best method to assess corporate climate and the company's management.

Placing a letterbox with '*Kaizen* ideas' written on it is the worst way to try and develop the acceptance of the *Kaizen* concept by the employees. The idea is destined to fail because it will involve a small number of operators, unless a previous effort is made to raise awareness, to organize, to train all the people upstream of the process. Downstream, the evaluation and implementation of the suggestion must be regularized.

Without such framework, the method might just produce generic suggestions, often with a provocative intent, which will undermine such an important initiative for lack of understanding and consolidation among the people involved.

Italians are well known for being innovative, for having a fertile imagination, for their ability to make do with what is at their disposal. Succeeding in the integration of these traits with the methods and the attention to detail that are typical of Japanese culture would unleash a great potential. 'To do a lot with little' has been an integral part of Toyota since the beginning and it is embedded in its DNA. The goal is to channel the creativity and experience of people to achieve something tangible for the company and for its growth.

We adopted several tools to maximize the benefits of this approach: a massive awareness and participation campaign was conducted for the entire workforce, training on the principles of TPS and on waste; from Japan, we imported and adapted tools such as the *Asaichi*, already outlined in previous chapters, quality control circles (QCC), which we will discuss later, and *YK-Y*. The latter is the acronym for the following Japanese terms:

- *Yarinikui* (**Y**): operation that is hard to perform.
- *Kizukai* (**K**): operation that is hard to perform which requires attention.
- *Yarinahoshi* (**Y**): operations to repeat and correct because they are hard to perform with a single action.

Figure 14.1 Collection point of operators' ideas

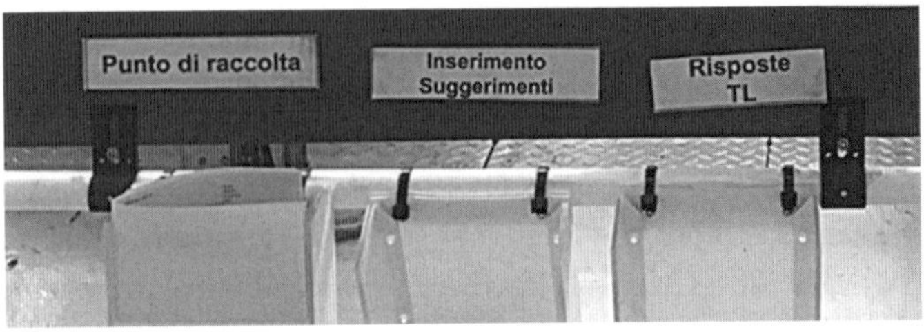

In each production line section managed by a team leader there is a board dedicated to ideas for improvement. There is also an ideas collection box (Figure 14.1) for all the operators working on that section of the line; they have a dedicated form they can use to describe the problem and suggest a possible solution. Their TL is always available to set up a meeting and discuss the idea or problem in question. Subsequently, the TL discusses the suggestion at a management meeting that sees the attendance of all the main corporate functions. The meeting will deliberate for a Go or a Stop, will decide which corporate function, if any, will be responsible for the implementation of the *Kaizen*, and it will plan its development and introduction. The TL must reply in writing to all the suggestions received and place the reply in a dedicated box, explaining the eventual reasons for a rejection. If the idea gets accepted, its implementation is communicated on the board, with planning and timeline, and pictures of a 'before' and 'after' its introduction.

Suggestions, information and all the material needed for the development of YK-Y are displayed right by the related area, in compliance with Toyota visual management philosophy. The board also includes the layout of the area, where the stations subject to improvement ideas are highlighted (Figure 14.2). A final classification into safety, quality, environment and production indicates the scope of the idea that was received.

Another system to promote *Kaizen* and the involvement of people is the QCC, quality control circle.[1] Born in Japan in the 1960s, the QCC has been adopted by several companies; Toyota was one of the first to apply it in 1967, seeing its potential in conjunction with TPS. Today QCCs are found in a large number of sectors, from manufacturing to public administration, because their goal is to set up a method that allows the small problems that arise during the performance of daily tasks at work to be solved.

The application of the QCC requires a substantial knowledge of TPS; it took us years to set the ideal conditions that allow their application in

Figure 14.2 YK-Y activities displayed

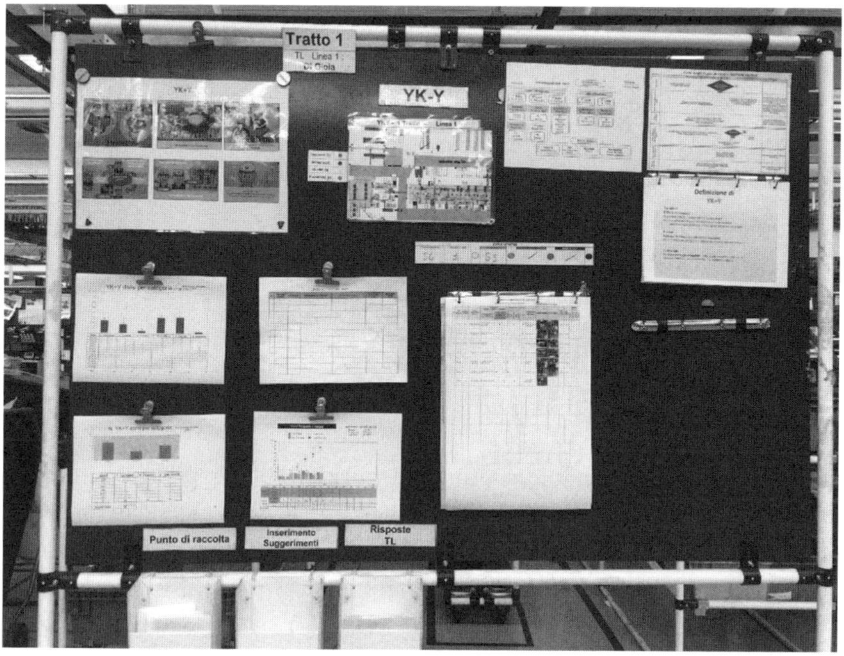

TMHMI. Its philosophy and the main quality control tools (see Chapter 11) are to be fully integrated in the corporate culture, and the company must create a framework that actively supports *Kaizen*. QCCs in Toyota see the involvement of the whole factory: lines are stopped every month for a one-hour meeting (or two meetings of 30 minutes each). Each QCC leader, who has been properly trained for the task, manages a small group of no more than 10 operators. The topics of the QCC, especially at the beginning, must be simple, and its solutions must be within the field of the team activities. This simple approach avoids discussing issues that are too complex, that involve engineering solutions or large investments: debating a problem that cannot be solved by the group itself is a demotivating factor for the team. Moreover, each team has six months to complete the QCC cycle and start a new one.

The success of this endeavour relies very strongly on the training of all the company's employees; it must be carefully planned, because not everyone knows how to apply problem-solving, understands graphs or how to break down issues.

The project of introducing QCC in TMHMI sees a gradual implementation for the first four years, starting from small groups and expanding slowly

Figure 14.3 Communication model in Toyota

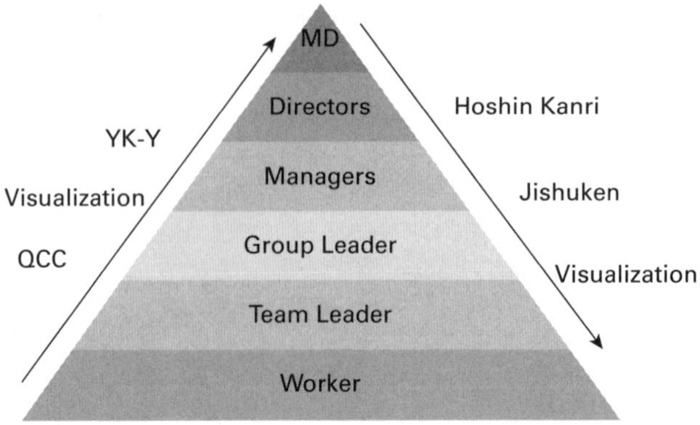

to the entire workforce. Interrupting the work of 400 people and setting up an organization that can manage more than 40 QCCs at the same time with total precision is not an easy task: it needs careful planning and an adequate amount of time and experience to succeed.

With the adoption of QCC, *Asaichi* and *YK-Y*, we fielded a set of tools capable of involving the whole workforce and promoting *Kaizen* activities, motivating people to be proactive in our constant search for the improvement of products and processes. The aim is to generate a sense of belonging, a shared vision: each one of us, regardless of role and hierarchy, can contribute to the success of the company and become a valuable asset in the long term. The widespread idea that there is a 'thinking' and an 'execution' side of the company, separating white- from blue-collar workers, must necessarily be changed.

The tools we adopted and integrated with *Hoshin Kanri* and *Jishuken*[2] set up a strong bi-directional communication system between top management and operators (Figure 14.3).

Notes

1 Quality circles are a method applied for problem-solving or quality improvement; small teams of workers meet with management to discuss and suggest potential improvements. Toyota uses them to solve quality problems, but they can be stretched to cover other targets: cost, delivery, environment, safety, moral.

2 *Jishuken* is an activity created to involve the managers in factory activities (*Genba*), with specific projects chosen by themselves.

15
The supply chain

Toyota sees its suppliers as partners, and as such it looks for agreements with them that can be as long as possible and beneficial for both parties. The level of service that is expected, the quality of the goods and a common culture that focuses all its attention on the customer (the customer first approach) are factors that strengthen the relationship over time; replacing a supplier is always a decision made after very careful considerations involving several different aspects. Instead of grabbing the short-lived opportunities offered by global markets, Toyota prefers to run its business and maintain its relations looking at the medium to long term, assessing each opportunity based on the TCO (total cost of ownership),[1] rather than the mere purchase price. It is fundamental to know and understand the supplier as an entity, assessing its financial strength, its culture, its approach towards business activities. It is also very important to evaluate its geographic location: it must be as close as possible to the plant that is going to use the parts it supplies. Although Toyota is a global conglomerate, these concepts can balance the need for economies of scale and the supply from international markets with the opportunities and the benefits offered by local businesses. European suppliers of Toyota Material Handling Europe cover about 90 per cent of the value of direct purchases.

It is important that the supply chain is located close to the production sites of the group, because it allows for a streamlined planning process such as the *Junjo*, and a logistics model that can organize one or more daily deliveries. We have often stated how lead time, or the time that lapses from the customer order to the delivery of the product, is a substantial factor in TMHE's business; short transport times contribute to its reduction.

TMHMI conducted a major rethink of the supply chain, reducing the number of direct suppliers down to about 200. This change focused the activities and the management of Tier 2[2] and 3 mainly on Tier 1 suppliers.[3]

We exploited the location of the supply chain to develop a *Kaizen* project in the inbound logistics area, aimed at optimizing the parts-picking activity in terms of transport costs and volumes retrieved. The location of each supplier has been visualized on a map with the use of flags, where equidistant circles are drawn from the Bologna factory. An analysis of the volumes of materials sent on a weekly basis has identified three different categories, which were assigned different coloured flags: green for high-volume suppliers, orange for medium volumes and pink for low volumes. The map highlighted the geographic areas where the suppliers were concentrated, while the flags showed weekly volumes of deliveries. This allowed us to identify and optimize a number of pickup routes that we called milk runs.[4]

Concentrating the collection of parts from multiple suppliers into a single route allowed us to split our requirements for materials into several journeys, reducing the stock that we had to keep in storage and maintaining low transportation costs.

Selecting a supplier implies several steps, where each element must be carefully assessed: financial solidity, certifications, the ability to develop new products while ensuring high quality standards and flawless management of its processes. An audit looks into all these elements and follows with related reports. The certifications that are required, apart from quality management systems ISO 9000, are ISO 14001 (the international standard specifying the requirements for an environmental management system) and OHSAS 18001 (for the deployment of a safety management system). It is likely that we will also start requiring ISO 50001, related to energy management. Obviously, the above mentioned certifications have already been awarded to all the production sites of our group.

Figure 15.1 Example of a milk run study

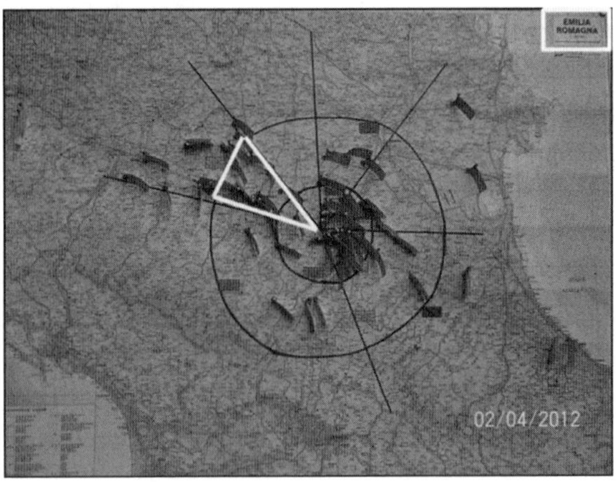

Each supplier undergoes a development plan, starting with the training of principles and methods of TPS, followed by a direct intervention on the activities, using increasingly complex tools, allowing for the continuous development of service, quality and efficiency levels (Figure 15.2).

The certification programme (Figure 15.3) monitors the supplier performance, with evaluation points given every quarter, based on five different parameters:

- Q: quality;
- C: costs;
- D: service level;
- E: certification ISO 14001;
- S: certification OHSAS 18001.

The score of each of them has a different influence on the final evaluation: more importance is given to the level of service and to the quality of the supplied product.

Certified suppliers can enjoy all the benefits offered by the group: for instance, they have priority over the future development of new products, and for potential supply of other plants of the group. The benefits are reduced for the suppliers that have been classified as 'qualified', while non-certified suppliers must improve their performance if they want to avoid the risk of being replaced.

The performance of the suppliers is displayed in the *Asaichi* area: it includes the chart of those that received the most non-conformity reports (NCRs) in the last few months, the suppliers that replied late to required corrective actions for faulty items, and finally the list of suppliers that did not reply to our communications. At times, not even Toyota, with its size and influence over the supply chain, is able to dictate its standards and requirements.

Our constant effort to improve corporate performance, optimizing costs and reducing stock levels, must not affect the supplier. Toyota knows that a supplier must have a financial margin that can support and expand the business; pressing a cost reduction by negotiating, or reducing the storage area while forcing the supplier to stock the materials, are tactics that can only bring benefits in the short term. We are fully aware of this, to the point of making our know-how available to the supply chain. We are the first experimenters of TPS, and our job is to reinforce the results we obtained and promote our approach with the supplier. The importance and value of

Figure 15.2 Structure of cooperation with the supply chain

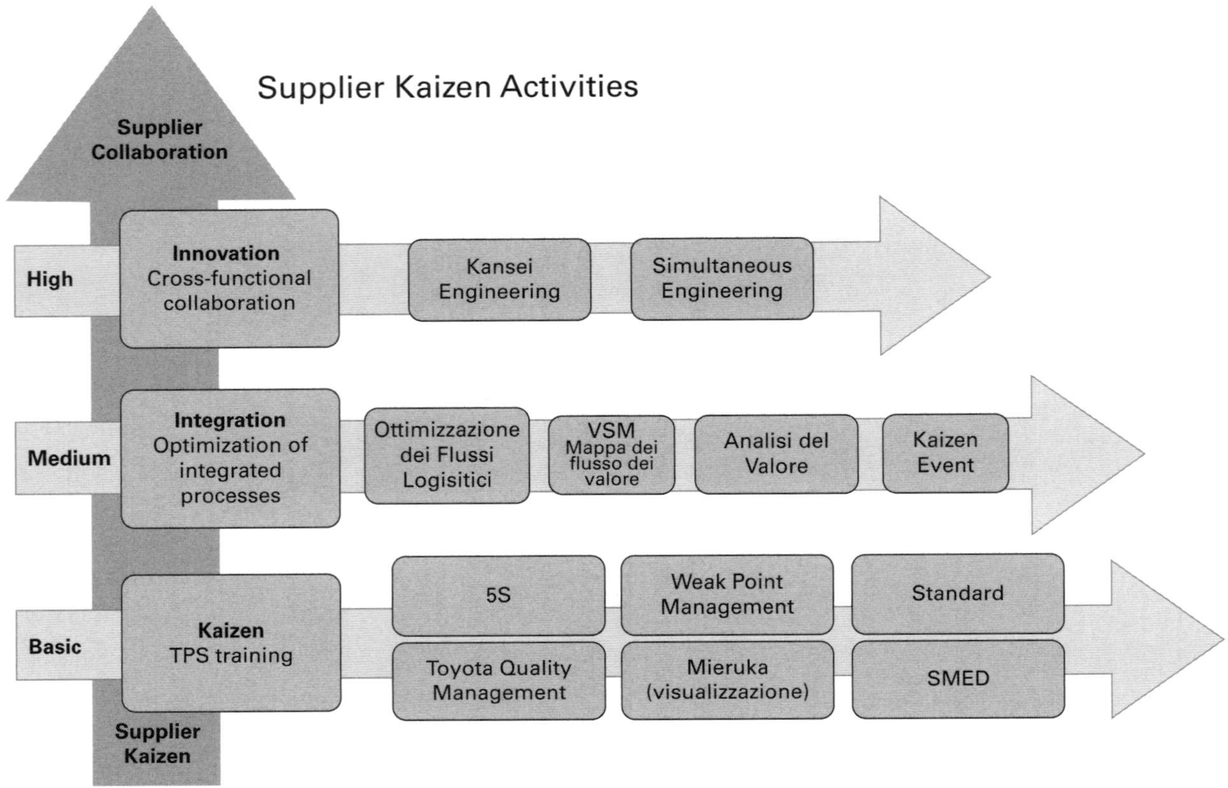

Figure 15.3 TMHMI certification programme

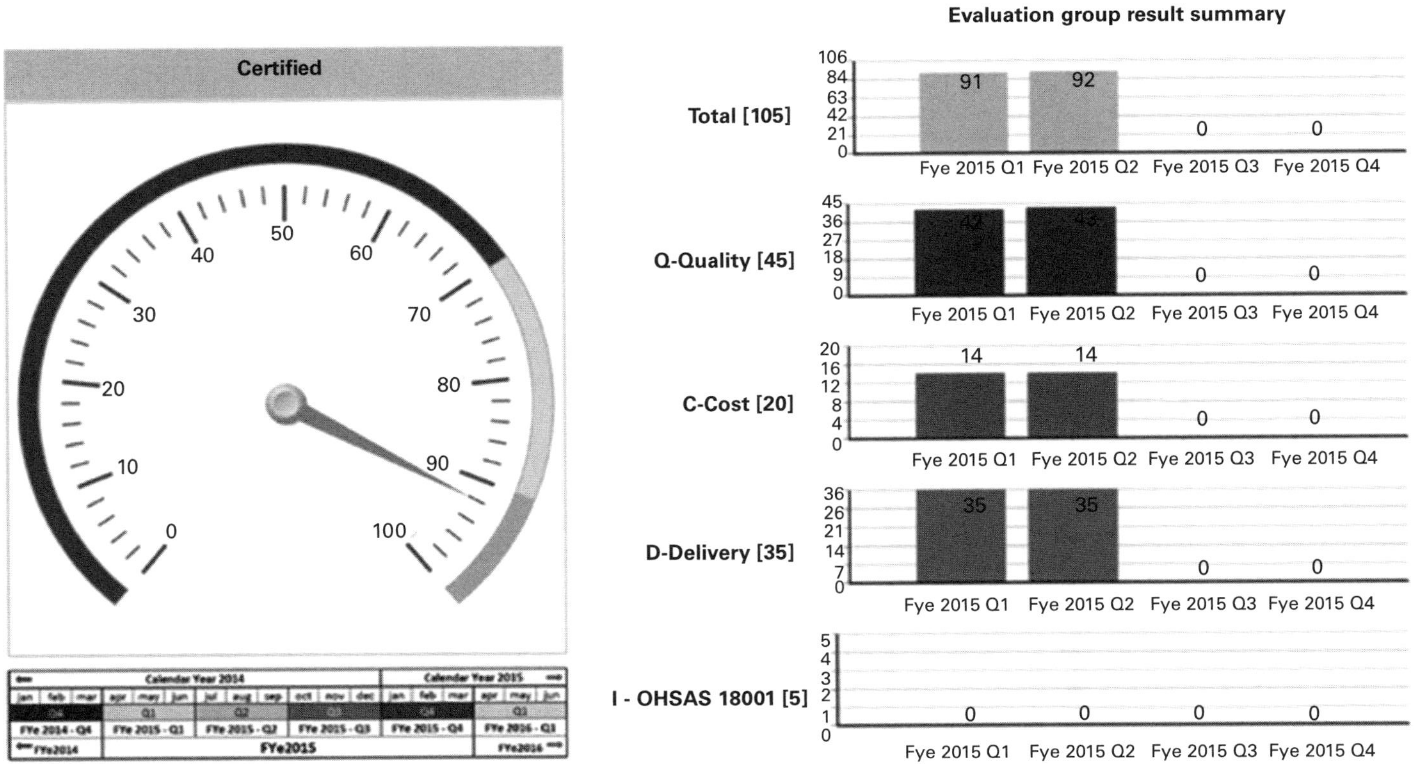

Toyota experience sometimes is not enough to convince the supplier to introduce TPS and modify its working methods. When we were given the opportunity to transfer them to our approach, the results have always been positive: the suppliers not only adopted TPS for the products that were destined to our sites, but they expanded it for their entire product range, obtaining great benefits overall.

Examples of implementation of TPS on TMHMI supply chain

Carpenfer S.p.A.

The company was founded in 1970 for the production of metallic carpentry objects. It supplies the main international producers of earth moving machinery, asphalt pavers and agricultural equipment (Figure 15.4).

The company supplies our Bologna plant with a wide range of forklift frames. Moreover, Carpenfer was the first supplier to which we applied *Junjo*.

Its first approach to TPS occurred in 2008, when TMHMI was launching a new, very relevant model for its future. At the time, the financial and economic crisis that hit globally after the collapse of Lehman Brothers had a massive, negative impact on the company results: their turnover shrank by more than 50 per cent in a single year. This situation, added to a sudden change of the company managers, caused a climate of uncertainty and confusion among the workforce. The activation of the TPS project, directly supported by the Bologna site, apart from the tangible results that it obtained on the products, allowed a crucial communication point between the new management and the operation areas to be established. Its typical bottom-up

Figure 15.4 Carpenfer plant in Reggiolo (RE)

approach and the implementation of simple tools such as blackboards and markers for the *Genchi Genbutsu* allowed management to work in close contact with the operators, building a relationship based on reciprocal trust. It represented an important opportunity for management to go to the workshop and check the problems and issues that were faced on a daily basis, sending at the same time an important signal of support to the workforce. The operators had the chance to communicate with the people responsible for corporate decisions, show them the challenges they were facing and suggest possible improvements. *Kaizen* brought great results in stock reduction, lead times and occupation of areas (Figure 15.5).

In addition to *Kaizen* projects, the company introduced the *Asaichi* and markedly improved the working environment by applying 2S. In order to involve all the employees and share the results that the company obtains, the management organizes a meeting twice a year, where the results are illustrated and discussed, and people have the opportunity to talk about critical issues and opportunities.

EnerSys Group

EnerSys is an American international company with a turnover that exceeds $2 billion. It is the world leader in energy storage solutions for industrial use. It owns several brands, and it owns development and production sites all over the world (Figure 15.6).

EnerSys is a historic supplier of Toyota; with the acquisition of CESAB by Toyota Material Handling, it strengthened its relationship with us, becoming one of the main suppliers for all the counterbalanced forklifts manufactured in Bologna. The battery is a crucial part for the trucks; its characteristics, coupled with the technical specifications indicated by the customer

Figure 15.5 Main results obtained with Kaizen activities

KAIZEN ACTIVITIES ON X540 FRAMES						
Description	**Units**	Previous figure	Current figure	Δ%	**Final target**	**Reduction (%)**
Number of frames in WIP	Days	8	3	–63%	1	–88%
Welding Lead Time	Days	26	20	–23%	12	–54%
Occupied area	m^2	112	42	–63%	14	–88%
Handling average	m	593	383	–35%	383	–35%

Figure 15.6 The Italian HQ of EnerSys

requirements, create a wide range of options. If we just take into consideration the brand number alone and the technical specs of the batteries available for the 48V model, we are looking at over 150 options from which the customer can choose. This range, associated with the cost of the product and the areas needed for storage, made us realize that *Junjo* would have brought several benefits. TMHMI therefore began an implementation project with EnerSys, aimed at developing a planning model that would allow a *Junjo* approach to the delivery of batteries to the plant. The right *Junjo* would contribute to a substantial reduction of the number of batteries in storage, consequently reducing the area needed and the total value of the stock in the warehouse.

Here is the experience of EnerSys Italia's CEO regarding this important transformation:

> At the time [June 2014] we were one of Toyota's main suppliers, with a 60 per cent share of their battery orders, and everything was going well. Then, one day, all of a sudden Toyota asks us to assess the possibility of delivering the goods on a daily basis. The possibility quickly became a requirement, so we began applying the new method. At the beginning everything looked extremely complicated and we had to face some serious issues. First of all, the fact that the goods were arriving from a plant located abroad, with two days' shipping time, would have required excellent coordination between production and logistics. In addition to the distance problem, it was important for us to saturate the vehicles and optimize the cost of transport.

Deliveries were already carried out twice a week, to ensure that Toyota had the parts needed for the following three days; now we had to study a new request and planning system that could guarantee that the goods would be ready three days before the required date, followed by their daily delivery.

After several internal meetings and conference calls with the people in charge of our production site in Poland, we created a Gantt diagram that clearly defined the structure of all internal projects, showing when to prepare the *Kanban* file, how much production we could guarantee to the customer, when to send the goods (Figure 15.7). We converted the *Junjo*/delivery plan into an Excel file and implemented several control checks to make sure that our order book was in line with TMHMI's. On Mondays, a further detailed analysis is carried out with the help of the delivery plan. However, the most important thing for us is the daily reception of the assembly sequence planned by Toyota, which allows us to see immediately any variations in quantity or dates and apply them to our management.

As far as the labelling of the goods is concerned, we made our IT department develop a customized solution that allows automatic uploading on our managerial system of the packing list data from the EDI system.[5] This reduced to a minimum the risk of mistakes in its preparation. This project required a considerable restructuring effort of our internal processes; however, it brought us several benefits, among them a Just In Time management of the ordered goods and of planning, WIP reduction and optimization of the vehicles' loads. These TPS-driven changes in the name of Lean manufacturing have brought significant overall improvements.

Bonfiglioli Riduttori

Bonfiglioli is another historic company based in Bologna, operating in several business sectors for an international market; among its divisions, TMHMI works with the mobile department, which develops and manufactures power transmission systems for self-propelled machines (Figure 15.8).

One of the main components that is produced for the Bologna plant is the transmission unit. The parts are usually made with a gear reducer connected to the forklift's wheel and to the electric motor of the truck. This is a strategic component, because it converts the power produced by the electric engines into motion. Its characteristics, costs and performance are key points from the very beginning of the development of a new product. For this reason, the supplier is actively involved with planning and creating a product

Figure 15.7 EnerSys planning system

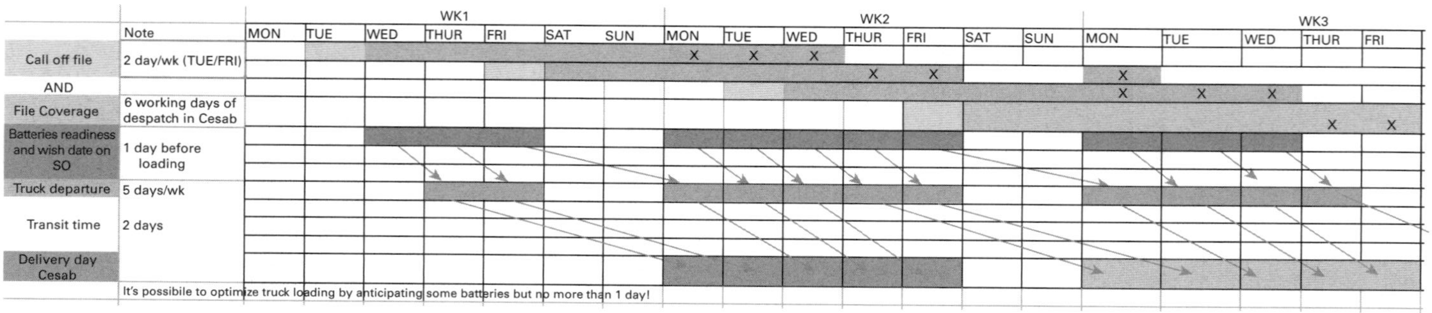

Figure 15.8 Bonfiglioli HQ in Italy

Figure 15.9 Transmission unit

that must comply with the quality standards set by Toyota Material Handling and with the performance of the new forklift.

Aside from their strategic relevance in terms of quality, cost and performance, the transmission units are fundamental, because without them it is not possible to begin the assembly of the forklift (Figure 15.9).

The fluctuations in TMHMI's final customer demands, and the increase in line capacity that the market required for the model equipped with the front axle supplied by Bonfiglioli, caused the deterioration of the level of service, worsening delivery times of this crucial element of the truck.

Figure 15.10 The steps of the development of the Bonfiglioli project

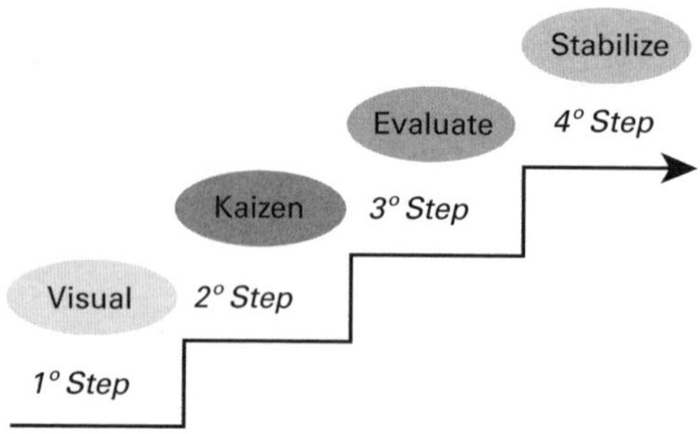

Following a number of meetings between the management of both companies, it was decided to create a common work group to spot the possible causes and find appropriate countermeasures. The project was split into four main points (Figure 15.10):

1 **Visual** (***Mieruka***)**:** analysis of the situation 'as is', identifying and visualizing with a checklist all the possible elements that affect delivery times.
2 ***Kaizen*****:** implementation of *Kaizen* activities to remove the main causes identified during step 1.
3 **Evaluate:** assessment and verification of the effectiveness of the corrective actions adopted.
4 **Stabilize:** monitoring over time the performance consistency.

We detected the most important causes for these problems, using a visual phase (Figure 15.11) carried out with the aid of a checklist indicating the main internal processes carried out by Bonfiglioli. Some of the issues were caused by delays in sub-supplies and a few bottlenecks in their internal processes. We shared and instructed the implementation of countermeasures and some *Kaizen* activities, with the addition of balancing buffers to compensate the difference in lead times and a careful monitoring of the compliance of delivery dates by the 13 main sub-suppliers involved in the production of the front axle.

The countermeasures and the *Kaizen* had an immediate positive effect on the service level offered by Bonfiglioli, which reached 100 per cent compliance with deadlines in just a few months (Figure 15.12).

Figure 15.11 The radar chart of the visual stage and the improvements achieved during the visits

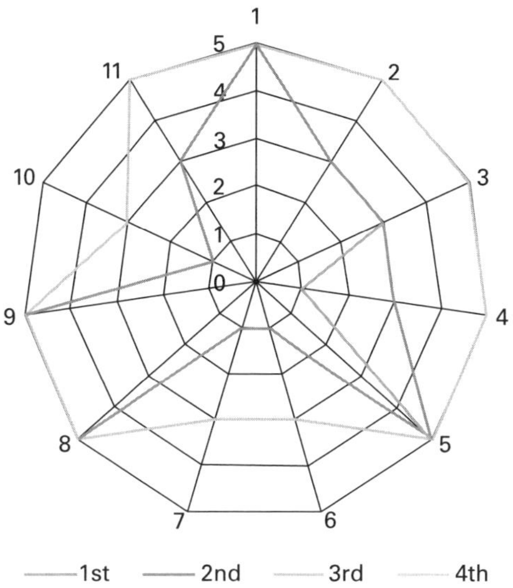

Figure 15.12 Delivery date deadlines by Bonfiglioli

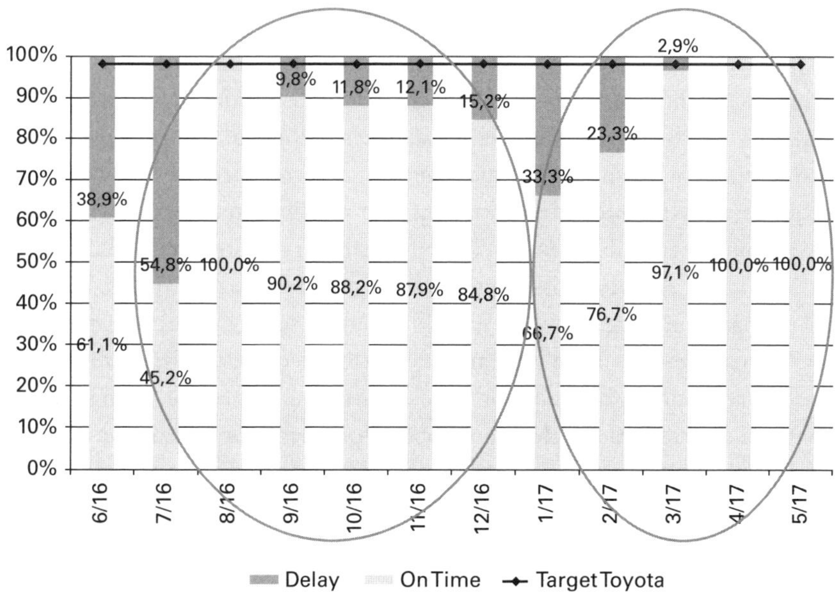

Figure 15.13 Würth Italia HQ in Egna (BZ)

Würth Italia

Gruppo Würth, with a turnover of over €13 billion and more than 73,000 employees, is the world leader in the distribution of fitting and assembly products and systems (Figure 15.13).

Active in the Italian market since 1963, Würth Italia has three logistics centres in Egna (BZ), where the HQ is located, Crespellano (BO) and Capena (Rome), allowing them to be as close as possible to their customers, among which is the Italian commercial branch of Toyota, Toyota Material Handling Italia: Würth is the sole supplier for all the small metal components used by the company workshops. The management of a considerable fleet of forklifts requires a well-equipped workshop, with qualified technicians capable of restoring all the trucks sent back from customers for repairs and reconfigurations in the shortest possible time. It is crucial that all the parts, including all the hardware, are readily available to reduce service times to a minimum.

The wide range of the fleet and of forklifts requires a disparate range of hardware, which is difficult and time-consuming to manage, with shortages that can interrupt the technicians' operations, who must wait for the order and delivery of the missing hardware. This problem was shared with the supplier, and together we assessed a number of solutions that could satisfy the needs of both companies:

- TMHIT needed to eliminate the risk of running out of the hardware needed to repair the forklifts sent for servicing;
- the supplier needed to be aware of the quantity of parts stocked by TMHIT to optimize planning and restocking.

Figure 15.14 Hardware shelves managed by Würth

The solution that was studied and agreed with Würth was to set up a hardware warehouse that included all the parts needed for servicing. The warehouse is inside the TMHIT workshop, but the management and ownership are Würth's. Basically, we created a 'consignment stock' model:[6] each time an operator finishes a box of a single hardware item, the box itself becomes a *Kanban* that Würth replaces within 48 hours. This supply method means that we stock and pay only for what we actually use, streamlining and simplifying the internal management of hardware. Würth has total control of its warehouse and can tweak its supplies based on actual usage by TMHIT (Figure 15.14).

Notes

1 TCO, total cost of ownership, is an approach developed by Gartner in 1987, used to calculate all the costs in the lifecycle of a piece of IT equipment: purchase, setup, management, repairs and disposal.
2 Tier 2: sub-suppliers who work for Tier 1 suppliers, or who manufacture parts that are not complex nor fundamental for the final product.
3 Tier 1: suppliers responsible for designing and developing main components.
4 The milk run is a round route that simplifies the distribution or pickup of goods. See Chapter 8.

5 EDI (electronic data interchange) is the data interchange between IT systems, using a dedicated channel and with a format that does not require any human intervention, save for a few elements. Its evolution allows it to be connected to the system through a simple Internet browser.

6 With consignment stock, the supplier sends the goods to the customer warehouse, but it remains the property of the supplier until the goods are picked by the customer.

16 *Monozukuri* and *Hitozukuri* (making people)

It is quite common to see the application of TPS on the tools that made it famous, like Just In Time, the *Kaizens* and all other systems and methods that we described in the previous chapters. The truth is that TPS is a radical cultural shift, and no tools can be really effective without the involvement of people, who must be at the core of this approach.

TPS can actually express its full potential only if it involves the operators who work in production. The creation of a 'friendly' environment, where people can perform at their best, is a goal of many companies. In Toyota, the best expression of belonging to a workplace is to be willing to contribute to its growth.

Japanese culture, with its past of isolation and military rule, has facilitated the application of these methods; the opportunity to manage the entire working cycles of individuals has allowed long-term projects and structured career plans right from the beginning of their presence in the company. It is also true that Toyota is a multinational conglomerate, part of the TICO group, parent company of Material Handling: more than 50 per cent of the workforce comes from different parts of the world, with the representation of a number of disparate cultures. Moreover, nowadays even the traditional loyalty of the Japanese employees is changing.

The acronym TPS can also be read as 'thinking people system', to highlight that it regards a cultural shift, not just a working method, where people play an active part in the development of the company.

This concept is summarized with the Japanese terms *Monozukuri* and *Hitozukuri* (Figure 16.1):

- *Monozukuri*: being able to manufacture quality products.
- *Hitozukuri*: being able to manufacture quality products through people.

Figure 16.1 *Monozukuri* and *Hitozukuri*

物 – Mono (Things)	
者 – Mono (People)	造り – Zukuri (Production)
人 – Hito (People)	

The ideograms used to write these terms have different meanings, which represent both the product and the people.

A company that puts people at the core has a history and a culture that make it more than a group of people sharing roles in a working relationship. In Toyota, right from the beginning, managers and leaders always tried to instil the values that can inspire how to run the company, both in its strategic choices and in its daily tasks. Such concepts derive by the principles written by the founder, Sakichi Toyoda, who summarized the values and philosophy that must inspire each worker. His five precepts are at the foundation of all the group's companies[1] and are among the main topics of management training. The training courses held by the Hazu Academy in Japan include activities dedicated to the founder and to the principles he wrote, from which derive all the group values (Figure 16.2).

Figure 16.2 The five precepts written by the founder, Sakichi Toyoda

TICO has written for them a modern interpretation, a fundamental guide for all employees.

1. **Always be faithful to your duties, thereby contributing to the company and to the overall good.**
   Focus on the same goal of contributing to the company, to society and to the planet, working with passion and commitment.
2. **Always be studious and creative, striving to stay ahead of the times.**
   Always search new, innovative ideas and excellent results, focus on competences and knowledge of everyone and constantly create value globally.
3. **Always be practical and avoid frivolousness.**
   Be honest, with your feet on the ground and with a direct approach, to understand the fundamental elements of everything.
4. **Always strive to build a homelike atmosphere at work that is warm and friendly.**
   Develop people and at the same time promote the creation of work groups that can face future challenges.
5. **Always have respect for spiritual matters, and remember to be grateful at all times.**
   Always feel grateful, remembering that our activities are, directly or indirectly, appreciated by people and society.

During our training we also visited the house where Sakichi Toyoda was born. It took us more than an hour by bus to reach our destination. Once there, we met an old man who, with the utmost kindness and welcoming spirit, conducted us through the rooms where the original furniture has been faithfully reproduced. He showed us with pride and respect, as if he was talking about a deity, the founder's inventions, kept in a small museum. This visit left a deep mark on me; it was as if I became part of a family, and it was important for me to know its story and identity, its heritage and ancestors. Today's Toyota, a cluster of companies that employs more than a million people all over the world, was born from the effort and dreams of a man whose curiosity and will to understand are condensed in one of his quotes: 'Open the window. There is a beautiful world out there' (Figure 16.3).

Each company of the group has adopted the precepts as reference values, which have been translated and spread in the languages used by all the business areas. Among these values, we can find the five elements of the Toyota Way (Figure 16.4):

Figure 16.3 'Open the window, there is a beautiful world out there' (Sakichi Toyoda)

Figure 16.4 The five values of the Toyota Way

1 *Genchi Genbutsu.* Already discussed in detail in the previous chapters, it is one of the most important precepts of Toyota's philosophy. Promoting contact with reality is the most effective way to have practical and efficient managers. One of the episodes that impressed me the most concerning *Genchi Genbutsu* is the strategy that Toyota adopted to enter into the North American car market. The Japanese market, where Toyota was leader at the time, is hugely different from the European and North American counterparts. The lack of space, an efficient long-range transport system and a different culture have led to the development of small-engine, compact cars with specific characteristics, mainly used for short trips. The chief engineer in charge of the development of a new

model destined for the American market, before drawing a single line on the board, rented different cars made by various competitors, and travelled for months through the roads of Canada and the USA. He soon realized that the American customer had different needs and expectations. A network of endless highways and bigger average drivers, alongside other cultural aspects, were not compatible with the cars destined for the internal Japanese market. This travelling experience, conducted to teach him how to design a product that could fit into this new, unknown market, allowed Toyota to offer cars that matched, and at times bettered, the competitors on American soil.

2 ***Kaizen***. One of the Japanese terms that became famous thanks to TPS and the Toyota Way, it represents another fundamental value for the group. Promoting *Kaizen* means promoting continuous improvement made of small steps, involving all the employees of the company.

3 ***Challenge***. A medium- to long-term vision is strategic for the growth of the company. If we promote a culture that encourages this approach, we can easily spot and exploit potential new opportunities. For instance, this approach made us the first to believe in hybrid cars.

   The *Hoshin Kanri* is one of the tools that the group uses to promote and share medium- and long-term strategies. The term *Hoshin* means 'direction, strategy', while *Kanri* means 'management, control'. At the beginning of every fiscal year, TICO's president introduces the medium- and long-term *Hoshin* for the group, and each division must prepare its own *Hoshin* accordingly. This process involves all the functions, including the GLs, and since everyone works towards a target contributing to the common goal, it has ensured a great level of efficiency.

4 ***Teamwork***. The intelligence of a single person will never be superior to the sum of the intelligence of several people. The team must be nurtured as a value, so that the company can grow. In addition to the team, Toyota must also take care of the individuals and of their personal growth.

5 ***Respect***. In order to cooperate effectively in a multicultural company, it is paramount to promote a climate of respect and collaboration. Japanese culture also respects age and hierarchy. It is not a coincidence that the bows made to greet superiors have a different inclination based on the job role of the superior being greeted. Respect must be shown to different people and cultures, to the individual, to society, to the environment and, first of all, to the customer. The customer is the real protagonist in Toyota: the customer is sacred. This concept derives from the term 'customer' itself, which the Japanese see as a deity (Figure 16.5):

- O (お): honour prefix;
- **Kyaku** (客): customer;
- **Sama** (様): special suffix, only for those people whom to show particular respect (illustrious, eminent).

Hiring excellent people, with a flawless school resumé and considerable professional experience, is not enough to create a successful company. The working environment and teamwork are crucial for their potential to be fully expressed, consequently creating value for the business. This concept is expressed by the Japanese term *Shokuba Ryoku*, which literally means 'a workplace that infuses energy and strength' (Figure 16.6).

Shokuba Ryoku is the energy that sometimes you feel when you visit a company, not only for the results that it has achieved, but also for the feeling of strength and team cohesion that permeates the workplace.

According to Toyota, three elements are required to generate this energy:

- **Acquired skill:** all the knowledge we acquired through education, specialization and experience.
- **Management skill:** the managerial ability to listen, implement and improve ideas, and the skills necessary to explain and instil the importance of what is being done. Some of the sectors are safety and environment, quality, maintenance and human resource development.
- **Thoughtful concern for others:** creating a workplace that is beneficial for the team.

Figure 16.5 Japanese for 'customer'

お 客 様

o Kyaku Sama

Figure 16.6 Meaning of *Shokuba Ryoku*

SHOKUBA RYOKU

(Workplace) (Energy and strength)

職場力

Figure 16.7 The concept of *Shokuba Ryoku*

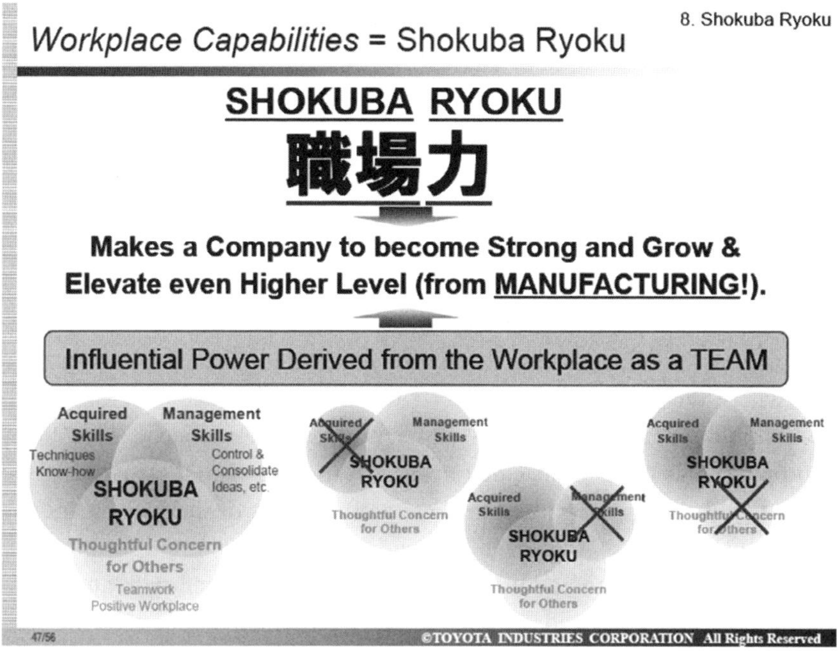

The blend of these elements forms an environment capable of transmitting energy and strength, where people can perform at their full potential (Figure 16.7).

Note

1 See the website of Toyota Industrial Corporation (www.toyota-industries.com (archived at https://perma.cc/P65R-GPPX)) for more information.

17
Toyota and union relations

A book about the industrial experience of TMHMI would not be complete without mentioning its relations with labour unions. Bologna, the area where the plant is located, has a tradition of strong, sometimes confrontational labour organizations; it was not the easiest of environments for us, in particular when we decided to carry out a substantial transformation of the factory, many of whose workers are members of the unions (mainly CGIL-FIOM) and are influenced by their decisions.

Union relations with Toyota Industries in Japan had their worst period in the 1970s, with strikes and demonstrations to complain about working conditions in the company. The events spawned the creation of an open channel with management that brought great changes, starting with a reference document called 'The relationship between company and employees, based on mutual trust and responsibilities'. The labour unions in Japan are arguably different from their Italian counterparts, which have procedures and dynamics that at times were difficult to translate and explain to our Eastern colleagues.

In TMHMI, as everywhere in Europe, union relations can have difficult times, which can be caused by external factors (such as the renewal of the National Collective Labour Contract), or by internal conflicts due to misunderstandings or failure to find an agreement.

One of the main topics, which caused an early misunderstanding of how TPS works, was the concept of 'more work'. A lack of information about the principles behind TPS erroneously transmitted the idea that TPS was created to make people work more. I still remember how the number of daily units produced was one of the main indicators that the unions used as a metric to measure the level of exertion. We started to implement a heavy training programme and we shared all the principles behind every project; the workforce slowly began to understand the concept of 'working better'. A clean workplace that complies with 5S, along with the organization brought by standardization, has contributed to the improvement of how work is organized, raising the output without causing a strain on people. For example, we closely studied the ergonomics of the plant and introduced a system for handling loads, such as zero gravity manipulators[1] and hoists, thus reducing manual handling as much as possible, even if the weight falls within the limits established by safety regulations. We created a 'listening board', where myself and HR were always available for discussions, and all the company departments gave total availability to listen to any issues reported by the unions, in particular regarding SPS[2] and maintenance. These changes eased the tensions during times of critical transformation for the company; the workforce's full involvement was crucial to ease our process.

The heavy use of visual information inside the plant helps to share information regarding good performances and weak points. For example, all the forklifts that do not pass the inspection tests are parked in a specific area and recorded on a dedicated blackboard, indicating the main issue and the department working on its correction. This indicator shows the WIPs[3] inside the plant; if the figure is increasing, it's the first signal that something is causing input (the number of trucks that begin assembly) and output (the number of trucks delivered to the finished products warehouse) to be badly balanced. Defective trucks are loaded on dedicated buffers, waiting for repairs. If the WIP exceeds the threshold level, repairs are carried out during overtime, even on Saturday when necessary, with the involvement of the operators in charge of repairs and final inspection tests. This simple KPI is an effective tool to display the problems we are facing and justify the overtime we need to apply: everyone can see how many trucks need servicing, what are the causes and what we decided to do about it.

The involvement of the employees through the direct and clear visualization of our projects in the related areas tends to avoid misunderstandings and the risk that important modifications to the layout or to work methods might be rejected by the operators. Nobody is willing to accept a change easily if they did not have any involvement in the change itself.

I believe that the *Yamazumi* chart, displayed in its reference area showing the data for the saturation level of each workstation compared with takt time, can be an ideal starting point, a set of data that can be assessed with the unions to spot any potential workstations with excessive workload. Moreover, showing the workload of all the production areas defeats the human inclination to think that our fellow neighbour does less work than we do.

Figure 17.1 Number of people we sent to Japan for training

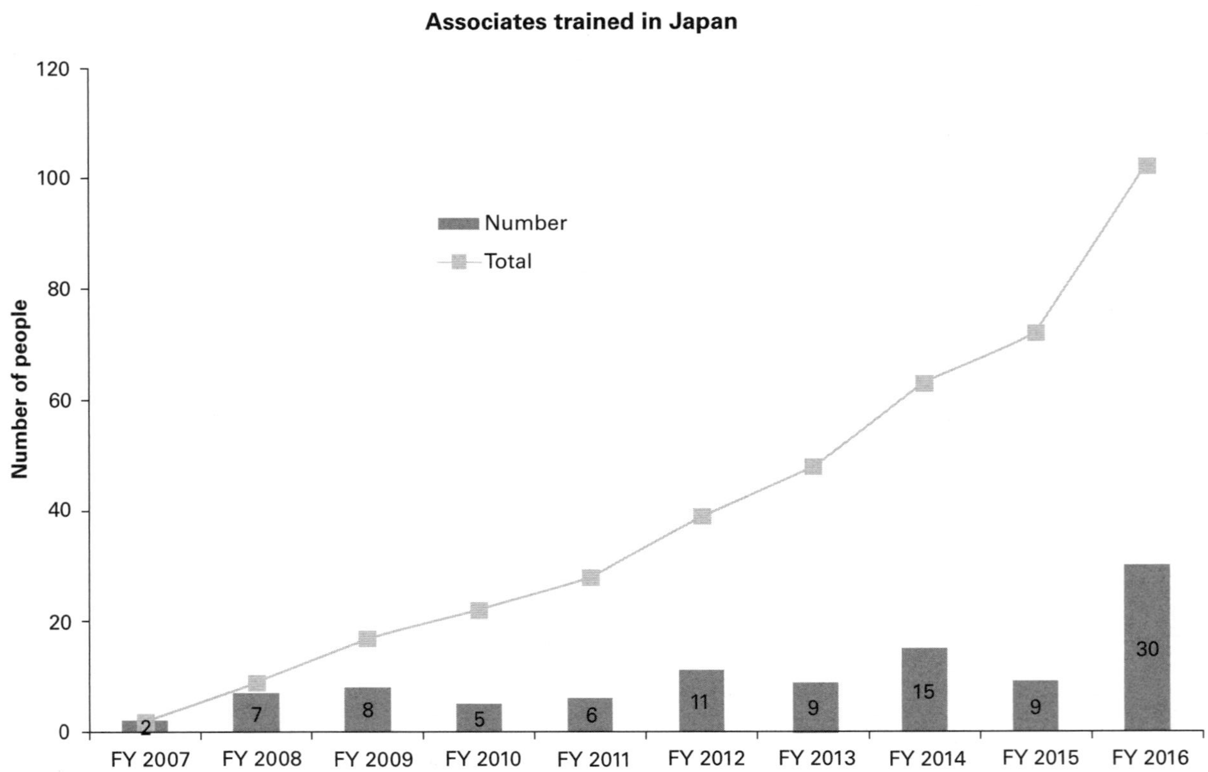

I cannot declare that we did not have any conflictual moments during these years of transformation. However, we all worked towards our goals, willing to find a common point that could be satisfactory for everyone. Our negotiations always prioritized the company and the people who worked in it, ensuring their future and security. In addition to improving their salaries with the introduction of production bonuses, we introduced other initiatives that improved welfare, such as a supplementary health insurance for all the employees and their families, and the opportunity to ask for paid leave for specific personal or family reasons. We paid particular attention to training: in addition to training carried out internally, we awarded trips to Japan for our workforce to be trained directly by the parent company (Figure 17.1). We increasingly adopted this method to share an important personal and professional experience with our employees.

Just as it happened at Toyota in Japan, we need to build a relationship based on trust and on respect of each other's responsibilities and functions, agreeing to work towards a common goal. Today, European manufacturing is feeling the strain caused by the insurgence of emerging markets; the only way to respond to this threat is to develop an excellent production and logistics system, keeping the company competitive and compensating higher labour costs with increased efficiency. The best action to ensure one's professional future is to give the biggest possible contribution to the growth of the company.

Notes

1 Gravity manipulators are devices for load-handling with an electronic system that can eliminate the weight of the handled good.

2 SPS: safety and prevention service.

3 WIP: work in progress, the number of trucks inside the plant excluding those in the finished products warehouse.

18
Kaizen and financial results

Another interpretation given to the acronym TPS is 'Toyota Profit System'. There are few business sectors that can decide the sale price based on the desired margin. In most sectors, especially in mature areas, the price is decided by the market: the margin for the company is strictly related to the product cost (Figure 18.1).

It is therefore crucial for company survival to maintain a virtuous cycle that constantly reduces the cost of the product. The reference market of Toyota Material Handling is mature; the reference price cannot be arbitrarily set by the company, because supply exceeds demand. Moreover, since the product is considered as a commodity by most customers, many deals involve short- or long-term rental agreements: price becomes even more significant when it's time to choose a brand. Constant reduction of the product cost is therefore a crucial element for business growth and for sustaining TMHE's position in the market.

At the beginning of each fiscal year, which for Toyota coincides with the Japanese year, from 1 April to 31 March, TICO's president introduces his *Hoshin Kanri*, from which those of each division will derive, a set of guidelines that trickles down to the group's plants. Each production site must develop its own *Hoshin Kanri*, a list of projects and activities that will support the group's target. Obviously, some of the company targets are financial, therefore the same must be applied to everyone. *Hoshin Kanri* and financial targets are intertwined.

TMHMI's plant in Bologna, developer and manufacturer of counterbalanced forklifts for the European market, is deeply involved in the activities aimed to reduce its product costs. In our case, a predominant portion of such costs is represented by external supplies; as far as we are concerned, it

Figure 18.1 Margin on products

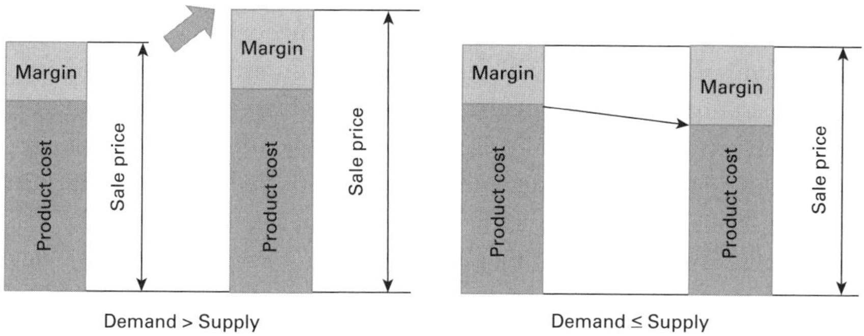

Figure 18.2 Trend of total manufacturing time, net of mix and models

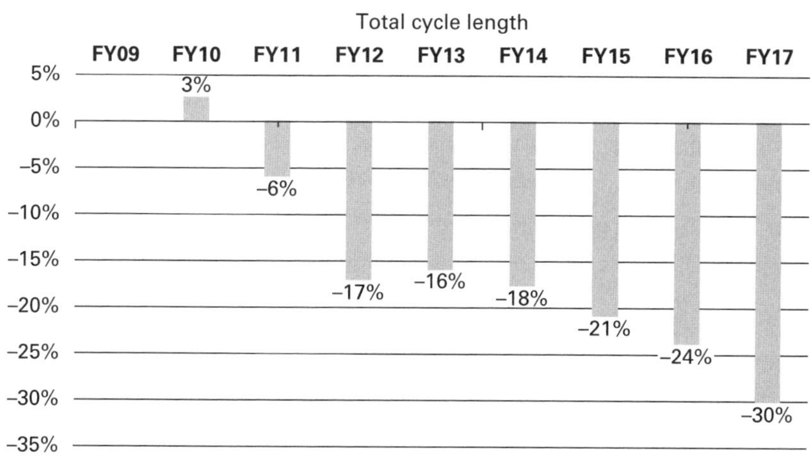

is crucial for us to renegotiate prices with the supply chain if we wish to hit our targets. In addition, we implement value analysis and value engineering activities[1] (Figure 18.2), jointly carried out with the supply chain, in order to eliminate potential non-added-value passages and reduce the cost of production. The operations department, receiver of a substantial share of the plant's investments, has the target to constantly reduce production costs; one of its main targets is to shorten total assembly times and transformation costs, thus lowering the cost per hour and consequently the total manufacturing cost.

The *Kaizens*, be them 'top-down' or 'bottom-up',[2] are the tools we use to develop the projects and initiatives allowing us to reach our desired targets.

Many activities have a tangible impact on production hours; we have split them into production hours and total working hours. Production hours indicates the time needed for the production of the forklift. It is usually related to a production cycle, with times allocated on the basis of an analysis of the operations that need to be carried out. Total working hours measures the total hours, direct and indirect, spent by all operations functions for the production of the truck. All indirect hours are reported by the operators in the time sheets, also indicating the cause and the activity carried out. This system allows us to categorize each activity performed inside the plant. There are three main categories:

- **Value-adding operations:** all the activities needed to give added value to the product. The assembly cycle and, in general, all direct hours worked on the product are part of this category. Direct hours, as explained in Chapter 12, are further classified in order to identify additional waste.
- **Reducible waste operations:** activities that are required but do not generate added value to the product. All the activities performed by services supporting production, such as material handling, quality control of incoming parts, office activities, etc, belong to this category.
- **Non-value-adding operations:** activities that create waste and need to be eliminated. Repairs and reworks are examples of activities belonging to this category.

Management control, taking into consideration the wide range of forklifts produced and the different production times based on models and options, publishes a monthly report based on allocated production times of each model and option, indicating the minimum number of hours that were needed to produce the mix and volume of the previous month. This value is the main reference point; when compared with the actual number of hours we used, we can establish our efficiency level. A 100 per cent efficiency level means that production used the exact number of forecasted hours; a lower percentage hints at inefficiencies that were required to be rectified, which translated into overtime or an increased number of direct operators. Using the production hours, all the direct and indirect hours used by the whole operations functions are verified, thus monitoring the general performance of the production area. The *Kaizens* we developed along the years have significantly reduced production times for each model, lowering production costs. The constant reduction of manufacturing times raised the production capacity of the lines, increasing the volumes without having to make big investments.

Figure 18.3 Volumes produced by the TMHMI plant

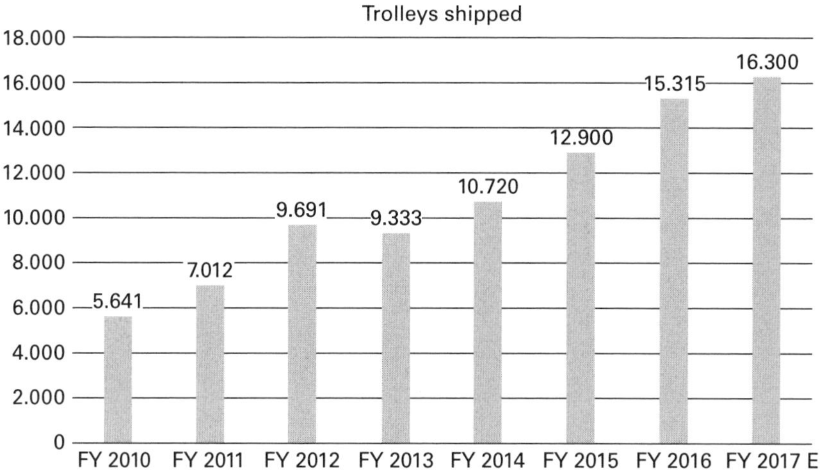

In fact, the production lines have remained substantially unchanged since 2001, but the *Kaizens* we applied to them succeeded in supporting the plant's growth.

A low level of investment helped minimize fixed amortization costs for new machines and equipment; such benefit, together with the increase in volumes, has positively contributed to the company's margins.

A further bonus deriving from a limited amount of investments is a lower break-even point (BEP),[3] which ensures that the company can post positive results even with decreasing sales. When the Lehman Crisis of 2008 hit our economy, the BEP was closely scrutinized by the parent company, to assess the level of flexibility of the group's plants.

The *Kaizens* that are suggested, aside from having an impact on either quality, delivery, environment, safety or moral, are often assessed for the impact they might have on production costs. Each *Kaizen* has a summary sheet that describes the activity, often through visual aids, with a 'before' and an 'after'. The final part of the document lists the savings that are achieved by applying the *Kaizen* and the time it will take for it to start paying back.

A key role here is played by management control, which, acting above board, must assess time and resources used, investments and the expected cost reduction, to ensure that the results are in line with the desired targets. Management control must be involved from the very beginning, to follow the development of the *Kaizen* and assess its overall impact. Often

Figure 18.4 Example of economic results obtained with a *Kaizen* activity

<table>
<tr><th colspan="8">KPI (Before Kaizen-After Kaizen)</th></tr>
<tr><th>QCDSE</th><th>ITEM/PROBLEM</th><th>Measurement Unit</th><th>BEFORE</th><th>AFTER</th><th>SAVING</th><td rowspan="3">Yearly effect</td><td rowspan="3">€ 54.483,00</td></tr>
<tr><td>C</td><td>WEL</td><td>Wel/workstation</td><td>0</td><td>ALL</td><td>22</td></tr>
<tr><td>C</td><td>Searching/Picking/Walking time for materials, Jigs and equipment</td><td>h/truck</td><td>5</td><td>2</td><td>3</td></tr>
<tr><td>C</td><td>Training time</td><td>Weeks</td><td>21</td><td>4</td><td>17</td><td rowspan="2">Payback (Months)</td><td rowspan="2">4,49</td></tr>
<tr><td>C</td><td>Efficiency/productivity</td><td>hours/truck</td><td>40</td><td>30</td><td>10</td></tr>
<tr><td>C</td><td>Assembly area reserved to ECOP</td><td>m²</td><td>186,55</td><td>78</td><td>82</td><td rowspan="3">Implemented</td><td rowspan="3">day month year
04 05 2016</td></tr>
<tr><td>D</td><td>Maximum number of trucks per week</td><td>trucks/week</td><td>2,5</td><td>7,8</td><td>5,3</td></tr>
<tr><td></td><td></td><td></td><td></td><td></td><td></td></tr>
</table>

the savings are indicated as 'potential', because the number of people involved is less than one, therefore it is necessary to assess the opportunity to combine the operator activities with others, up to saturation. There are occasions when the *Kaizen* frees resources that can be used for indirect activities, thus saving working hours. Other times, the real effect of the *Kaizen* on the product cost is reflected in a change of takt time that can further optimize the workstations, raising volume without increasing or decreasing the workforce.

Management control is a crucial, active part of the processes that involve TPS; it is fundamental that it integrates all the principles, adapting the control and organization systems to its philosophy. This transformation must be rooted deeply into the core of the company's activities. If management control, for example, develops analysis methods based on the minimum production batch, it might assess the product cost based uniquely on the saturation level of the plant; production volumes would be consequently set on capacity, not customer orders. In this case, the plant efficiency would not bring any benefits, because saturating the plant without a real demand from the market means more units to keep in storage; increased storage means more dedicated space, risk of obsolescence, potential damage to the stored units, company capitals frozen in the warehouse and other hidden costs. This is why overproduction is seen as the father of all *Muda*.

Notes

1 Value engineering is aimed at new products, while value analysis is concerned with existing products. It involves a current product being analysed and evaluated by a team, with the aim of reducing costs and/or improving the product's functions (www.advice-manufacturing.com (archived at https://perma.cc/GVR4-62QL)).

2 'Top-down' is a target process that starts from management and reaches all underlying functions; 'bottom-up' starts from the operational functions of the company and goes in the opposite direction.

3 In business economy, the break-even point is the point at which total costs and total revenues are equal. There is no net loss or gain, and one has 'broken even'. All costs that must be paid are paid, and there is neither profit nor loss (Wikipedia).

19
TPS and Industry 4.0

To complete the first part of this book, I would like to share with the readers some considerations regarding Industry 4.0.[1] This important industrial revolution, which is still difficult to outline in its entirety, sees digitalization and technology as the dominating elements of the whole supply chain process. Let's begin by recalling the previous industrial revolutions, which were characterized by:

- the introduction of mechanics, using steam for power generation;
- mass production, with the introduction of assembly lines (by Ford) and electricity;
- computers and automation;
- cyber physical systems (CPS).

Industry 4.0 introduces the concept of CPS, physical systems connected with IT structures that can interact and interconnect with each other. This concept is also defined as the 'Internet of Things'.

I have attended a number of lectures and conferences that depicted the future as a totally automated plant, where people are helped by robots in the more complex operations, and the connections among machines allow the execution of the entire production process, from the customer order to its delivery. I attended a conference that discussed this topic organized by a leading company of the sector, where its representatives listed the benefits that derived from the application of Industry 4.0; when they claimed that they could improve forecast by 85 per cent I couldn't help but smile, because at the same time, although thousands of forecasts made by the most reputable companies in the world went the opposite way, Donald Trump was being elected 45th president of the USA.

Industry 4.0 will definitely be a revolution, or better, an evolution of production systems, bringing the benefits of digitalization into operations. The real application of augmented reality, in areas such as logistics, and the installation of robots that can interact with people, will bring great changes and improvements of the industrial processes. However, while Industry 4.0 depicts a future made of 'machines', I believe that people will remain the core element of process optimization; Toyota calls it 'human touch'.

First of all, we need to focus on some mass phenomena: many businesses thought that the installation of a state-of-the-art ERP (enterprise resource planning) was indispensable to ensure business success, the same reasoning that concluded that transferring production to East Asia or to low-cost countries was the right alternative to investing for the improvement of their processes, solving at the same time the issue of high labour costs. Toyota in itself does not have a sophisticated ERP; the underlying philosophy is that the system must adapt to the evolutions brought by people, not the other way around, and therefore it must be adaptable and capable of supporting the evolution and management of new processes. The relocation of manufacturing activities in other countries has been inevitable for some sectors to ensure the survival of the business, but for many the choice was devastating. Some of the industries realized too late that there were many hidden costs that did not surface during the planning stage. This, along with other reasons, is causing the re-shoring of many activities, or the relocation of manufacturing activities directly in their reference market.

TPS is a cultural shift that brings people to think differently, that helps them find the *Muda* (waste). There will be waste even in a digitalized world; actually, the risk is to digitalize waste too, if the processes are not optimized to begin with.

Toyota is studying solutions that involve digitalization, always maintaining a gradual approach, implementing only the technology that can bring real benefits with reasonable amortization periods. For example, we began in our Ferrara plant some pilot studies on predictive maintenance, using and studying the analysis on vibrations and on the temperature sensors placed next to the moving parts in the painting department. The introduction of TPM (total productive maintenance) allowed an effective analysis of the real causes of breakages, which pointed to the critical components under stress: we have installed sensors on those components that will monitor the process and help us prevent future breakages.

One of the sectors that will be most affected by Industry 4.0 will be logistics. IT support will allow for maximum flexibility and quick order processing, starting from the order confirmation by the customer, right to shipping and delivery.

Figure 19.1 Augmented reality in the warehouse

A few years ago, I conducted a study to assess the potential advantages that augmented reality could bring in the preparation of kits and in the supermarket activities of the Bologna plant. Smart glasses supported the operator with information superimposed over the lenses (Figure 19.1). An indoor navigation system could optimize the route, showing the user the fastest way to reach the aisles where the parts were stored, while a localization system would correctly detect the code that needed to be picked. The results highlighted a notable reduction of picking errors, and an increase in efficiency brought by the optimization of the routes.

My opinion is that TPS can support any kind of tech revolution; its simplicity and effectiveness can make visible what will eventually be hidden behind complex analysis and processes of the digital world. To confirm this belief in true efficiency regardless of the approach, Toyota still uses a paper *Kanban*: the passage to a digital format has been deemed as a potential risk that might cause confusion and mistakes by those using it.

This is the reason why the merge between the Toyota Production System and Industry 4.0 can be one of the most successful moves in order to improve the businesses of the future.

Note

1 The term Industry 4.0 indicates a trend of industrial automation that integrates new production technologies to improve working conditions, raise productivity and product quality of the plants. Industry 4.0 takes its name from the industrial plan of the German government published in 2011 and implemented in 2013, which included investments in infrastructures, schools, energy, research centres and businesses, to modernize the German production system and bring manufacturing back to a leading global position.

Part Two
The operational implementation of the Toyota Way in the commercial and sales processes: the Toyota Material Handling Italia Convention

by Leonardo Salcerini and Danilo Verga

20 Foreword

The Toyota Academy case

The start date of the dealers' yearly meeting was approaching: the Hotel Atlantic in Riccione hosted the event from 9 to 11 October 2009. Leonardo Salcerini, managing director of Toyota Material Handling Italia, was in the company of other senior collaborators of Toyota Academy[1] in their offices, alongside all the other employees, to plan, prepare and share what would be the topics of the imminent meeting.

Leonardo Salcerini was quite worried. It was the first time that he gathered together both indirect BT CESAB and Toyota sales networks, which had been fierce rivals until then. Unsurprisingly, he was taking a lot of care in the preparation of the convention, where he would have announced the merger of the networks.

The presence of new collaborators, previously associated with Toyota Academy, required the implementation of a new process of vision-sharing and communication, something capable of illustrating the past, present and future of the entire company, on top of setting up the guidelines for the organization of the workshop. In addition to the description of the guidelines for Toyota Material Handling Italia commercial strategies, they would have introduced and shared Toyota values, their founding elements, the specific declination of the guidelines that outline the Toyota Way and its vision, added with the education and consulting activities offered by Toyota Academy.

Preparing the contents for the convention: the Toyota Academy – scope

Preparing the contents for the workshop required setting up a specific and detailed 'strategic guide dossier', consisting of various documents, of which these pages will provide a short outline.

This dossier, called *Toyota Academy – Scope*, in addition to including the specific documents regarding Toyota Academy's vision, contained:

1 a synthesis of the main training and educational initiatives dedicated to the dealer network of Toyota Material Handling Italia;
2 a synthesis of the main training and educational initiatives aimed at the old and new customers of Toyota Material Handling Italia;
3 The outline of several consultancy projects on advanced business management, values and implementation strategies of TPS offered by Toyota Academy, dedicated to:
    a. entrepreneurs and managers of small and medium-sized companies;
    b. multinational companies and organizations;
    c. supply and business networks, both local and international;
    d. public organizations;
4 a rich series of documents, projects, studies and cases that described in detail strategies, methods, tactics and development and transformation processes of direct and indirect sales networks, multi-branding strategies, successful best practices regarding the politics, and the commercial and sales choices that brought growth and success to the business management activities implemented by Toyota.

The series of documents that were prepared for the workshop focused on the development processes that had occurred in the recent past, which outlined the development of commercial politics and strategies of Toyota Material Handling Italia, and needed to be shared with all the participants of Toyota Academy's workshop.

An overview of the documents that were introduced are described in this paragraph and in the following chapters, documents that include:

- the intervention of the dealership 'Giovanni Speranza';
- the evolution of the Italian market for forklifts;
- the introduction and synthesis of the intervention of Leonardo Salcerini at the convention, regarding: the actual situation of the Italian market for forklifts; the issues related to the merger of the indirect sales networks; the application of Toyota values and the creation of a new strategic and organizational approach to market and sales;
- the interpretation of the Toyota Way and its contextualization in a mid-sized Italian business, and the new A.C.C.E.S.S. model of a collaborative approach to marketing and sales.

Note

1 Toyota Academy – Training and Consulting Division is an initiative by Toyota Material Handling Italia, in order to develop Toyota's business and management culture, Lean organization principles, the Toyota Way values, processes, methods, innovations and declinations of TPS, with publishing projects and educative events, initiatives and consulting activities. The aim of Toyota Academy is to develop entrepreneurial and managerial competences in people both within and outside the world of Toyota Material Handling.

21 The point of view of the dealers

Speech by Giovanni Speranza[1] at the Toyota Material Handling Italia Convention

Good morning everyone, I am Giovanni Speranza, owner of the dealership Superlift 4.0.

I spent a long time thinking and preparing this intervention: for myself and for all the dealers, this meeting has a fundamental strategic importance. What will be decided is going to deeply affect our business organizations and future activities.

What is the main reason for us being here today? We must outline the future of the sales networks BT CESAB and Toyota. We were competitors until yesterday: we often fought against each other to entice, conquer and keep our customers, which are our true and indispensable asset. Today it's time for us to rethink our strategies to approach the customer, to reset and reorganize all our work.

This occasion sees all of us dealers together, for the first time, face to face. Today's meeting celebrates the merger of the sales networks BT CESAB and Toyota, until yesterday open and fierce competitors. Today, our main goal is to define new, clear and convincing strategies, as economically viable and as harmonious as possible.

The most important goal is to agree on new behaviours, new organizations, new approaches to the customer and to the market opportunities in our reference areas.

We will succeed if we can agree on new solutions of collaborative intervention that are profitable for us all. We will have to avoid the bloody and devastating price wars, the discounts and the special offers that erode our bottom lines. We need to act with rationality and method, programming a joint course of action that can help us recover our lost market share.

We must, in a nutshell, 'get our act together', setting up an effective and successful organization, so that we can attract new customers and aggressively chase any market opportunities.

This is our problem today, and we must solve it together; we must get our act together and recover, achieve the profitability we need and ensure that our businesses can flourish harmoniously.

We also must find a strategy to avoid losing further sales opportunities and to recover the market share we have lost!

What do we actually have to do?

We must make available forklifts to hire in the short, medium and long term; we must change the way we work among us and with our territory manager; we must avoid cannibalizing brands and business opportunities, enhancing and differentiating our basic range, integrating it with the innovative services offered by Toyota (and by the dealers) and with our specific offerings.

Most important, and forgive me for repeating this, we must pay attention not to start useless and devastating price wars among ourselves!

Our professionalism will play a crucial role to conduct a competitive but fair contest, something that can create profits for each one of us and value for Toyota Italia and our customers.

The current market evolution will drive us towards a new assessment and evaluation of the intangible components of our offerings. We must also be ready to manage the historical shift that is changing the traditional sale of trucks into the short- and medium-term rental agreements, in order to be able to offer further services, accessories and spare parts.

Needs and expectations of more technology, quality, reliability and attention to safety

I would like to underline the importance that our reputation has on our offerings, a reputation built by following the typical Japanese standards for technology, quality, reliability, innovation and attention to safety, albeit with different characteristics for each brand (BT CESAB and Toyota).

As far as our overall offer is concerned, we need to consider that the sales networks have different needs pertaining to the quality of the range for sale.

In general, we constitute two indirect sales networks belonging to two different companies that, until today, have developed, promoted and offered two ranges of forklifts that are partially different, whose single and specific productions and models require individual modifications and attention for future developments.

With reference to quality, some CESAB products need a quality and technology upgrade that can make them more reliable and closer to the competitors' specifications; they also need to be more compliant with Japanese quality standards. On the other side, the products offered by Toyota need to have a more localized approach, more 'made in Italy', more flexible when it comes to production.

CESAB offers us a wide and flexible range; each forklift is unique in its genre. CESAB is a brand characterized by extreme flexibility, although there are shortcomings when it comes to quality. Toyota, conversely, offers a range of products that are of excellent quality; however, its offerings tend to be more limited, less flexible, suffering from range limitations.

I wish that the Toyota brand could achieve more flexibility, a range with a wider set of options, covering for and satisfying the needs of the final users.

If we look at the issues caused by safety, Toyota's production is very careful on the matter: it has introduced an exclusive, patented system called 'SAS' (System of Active Stability to prevent forklift overturning). I hope that such implementations and improvements can be extended to the CESAB range. In other words, I hope that the CESAB offerings can soon offer the same features adopted by Toyota to improve its operational safety levels.

Another critical aspect of the features offered by the forklifts is **technology**.

We all know that the technology of our machines goes across the board and involves all the aspects that define the value of our products. The aspects that are more coveted when we think about technological improvements are known. They have to do with efficiency and productivity, elements such as the speed of lifting and translation.

Nowadays ergonomics plays a significant role in technology, typically found in several features that are much valued and sought after. Ergonomics can acquire different value, ranging from air conditioning to improved access and comfort in the cabin. However, the technology that has most appeal today is related to performance. Customers look for flexible, versatile and high-performance products, with technical specifications that are becoming increasingly advanced and tailor-made. For this reason, I hope for the technological advancements that can improve performances constantly, therefore raising the desirability level of CESAB forklifts.

Us dealers expect to see a technological progress of the new CESAB models, aimed at improving and ensuring more advanced and competitive performance levels. We want CESAB trucks in line with Japanese standards, reliable through quality. My only wish regarding Toyota's offerings, again, is to be able to sell a product with more flexibility and options.

Integrating BT CESAB and Toyota indirect sales networks: dealership issues and the need for new distribution and sales policies

If, on one hand, the proactive involvement of the company to ensure a product upgrade is surely a positive factor, I also wish for management and sales organizations to develop the correct strategies and the right, innovative sales policies that can guarantee the operational continuum of our dealerships after the unification of the networks. I am asking for clear answers, solutions and continuity, instructions and guidelines; I wish for clear rules and interventions that can mitigate the issues that will certainly occur with the integration of the networks.

The anxiety in us dealers grows with the uncertain answers given to key questions related to the continuation of our agent licences, and to the future financial dynamics of our dealerships, which furthermore must now share the current complex and uncertain market opportunities with other colleagues, people who were competitors just the day before. The questions that need to be answered, for which we expect a positive and tangible approach starting from today, are the following:

- What will happen to the licences of the dealerships?
- Will all the licences be confirmed, or will there be cancellations?
- Will the dealerships manage to coexist and achieve positive financial and economic results although they serve the same areas?
- How will the integration of the sales networks be conducted in terms of company management?
- Who will monitor and supervise the relationships among dealerships located in the same area, and how will this be done?
- How will the likely issues and conflicts that will occur between dealers, previously belonging to different networks and promoting separate brands, be regulated and resolved?

- What will happen if another dealership offers the same product and targets one of my established customers?
- How will the promotional activities towards potential and existing customers of each dealership be regulated?
- How will Toyota behave towards key accounts located in an area covered by a dealership?

We need to define immediately clear rules of behaviour among dealers, and to identify politics, programmes and initiatives that not only must avoid useless overlaps, but that are also aimed at recovering the sales that we have lost because of the contraction of the economy due to the current crisis. A discriminating factor in market development will certainly be the limited financial resources of the dealers, which alongside the current credit crunch will significantly affect our potential growth.

It will be important to split the market into appropriate segments, mapping our customers and applying a market approach that will help avoid useless and costly competition between dealers, wasting energy on customers that we already acquired and repeatedly served.

We of course need, as I have already said, to update our offer; we need new products, manufactured following the suggestions I have listed earlier; we need units that can be hired and re-hired to several customers with different needs.

In order not to lose market opportunities and to reply to strong offers by the competitors, we need forklifts that are easy to configure and that can be adapted to the requirements of multiple customers. We also need a manufacturer that pays attention to those needs and can satisfy them in a timely manner.

If we can soon rely on speed and flexibility of manufacturing, and on reliable products, we will be able to offer our customers a range that fits their requirements, integrating it with a series of support and maintenance solutions that we can supply in a competitive, qualified and timely way, with no delays or further disruptions.

I am still worried about how we will manage to create value in our operations, guaranteeing our profits and respecting each other; we must agree and follow an ethical, clear and fair modus operandi, respectful of the work that each one of us has done, of the steps that we took to safeguard our organizations; we need to recover and grow without obstructing each other, finding new ways to cooperate, strengthening our professionalism and safeguarding our customer base.

Last but not least, we will have to trust the transformation of the factory, the innovations, the improvements of our products and the new commercial strategies that I hope will be guaranteed in the future by the management of Toyota Material Handling Italia.

If we reach these goals, if we are able to perform all this, we will easily be in a primary position to exploit market opportunities, strengthening our relationship with customers and guaranteeing our growth, profitability and a future full of successes and great achievements for everyone.

Thank you very much for your attention.

Note

1 This intervention was carried out by a well-known and reputable dealer. Some of the content and references have been omitted. The name of the dealer and dealership have been modified to safeguard confidential information.

22 The evolution of the Italian market for forklifts

Table 22.1 Sale of forklifts in the 2007–2009 period

2007	Year of maximum sales expansion in Italy: 46,076 units sold.
2008	Sharp decline in sales.
2009	At the beginning of 2009, the Italian forklift market showed issues that required action by the management of Toyota Material Handling Italia. The market almost halved if compared with two years previous, recording sales of only 24,427 units.

Analysis of the variation in demand for forklifts in Italy during the 2007–16 period

The analysis of the dynamics of the demand for forklifts by industry, large retailers and logistics companies during the 2007–16 period shows how the market started shrinking right from the beginning of the economic crisis, which began in 2008 and affected all businesses. Right from the beginning of the crisis, the volumes started to decrease, in particular sales of counterbalanced forklifts, which are mainly destined for the industrial sector;[1] on the other hand, large warehouse forklifts used by retail and logistics did not suffer as much, showing a growth pattern during the period under consideration.

Performance of the Italian market for counterbalanced forklifts

As previously noted, the performance of the forklifts market during the period in question has its peak during 2007, recording total sales of 22,673 frontal forklifts. The decrease in demand begins in 2008[2] and sees a total reduction of 16.6 per cent of the market for industrial forklifts compared with the previous year. Such decrease worsens in 2009, recording –59.5 per cent if compared with 2007's sales.

In particular, the sub-group of electric forklifts, after its peak of 16,525 units sold in 2007, sees sales the following year stop at 13,450 units (–18.6 per cent) and crash in 2009, with only 6,540 trucks sold, posting a tragic –60.4 per cent if compared with 2007.

The same trend is followed by frontal counterbalanced forklifts and by the sub-segment of diesel and gas-powered trucks, which sold 6,148 units during its peak year, sliding to 5,458 in 2008 (–11.2 per cent) and crashing in 2009 with 2,643 sales, which means –57 per cent in only two years.

Performance of the Italian market for forklifts destined for large-scale retailers and logistics

The same trend can be seen in the evolution of the Italian market for warehouse forklifts used by large retailers and logistics.

As we can see by the figures shown in Appendix A,[3] this sector recorded 23,403 units sold in 2007, marking the first slide in 2008 with 20,985 sales (–10.03 per cent). The following year saw another decrease, putting the total contraction of the market at –34.09 per cent for the period.

Long-term analysis of the Italian market for forklifts (2007–16)

In order to fully understand the complexity of the issues related to the change in dynamics of the forklift market in Italy, it is useful to extend the analysis over the long term and extend our observation from 2008 to 2016:[4]

a The analysis of the evolution of demand in the Italian market for counterbalanced forklifts in the 2007–16 period shows that:

- 2007 recorded the highest level of demand for industrial frontal forklifts, recording 22,673 sales;
- 2009 marked a significant decrease, with –59.5 per cent in two years;[5]
- 2016 shows the results of a recovery of the total demand for counterbalanced forklifts. During the first years of the crisis (2008, 2009) we recorded a total decrease of –20.07 per cent[6] compared with the peak volumes reached in 2007. We must take into consideration that, even with the recovery we experienced in the long term, the demand for frontal forklifts has never matched 2007 levels.

b The analysis of the evolution of the demand in the Italian market for forklifts destined for large retailers and logistics during the 2007–16 economic crisis shows that:

- 2007 recorded a total of 23,403 trucks sold;
- 2009 saw the demand shrinking, with 15,244 sales, marking a total –34.09 per cent decrease from 2007;[7]
- during the following periods, the trend shows a marked inversion: sales in 2016 reach 28,296 units, a +85.62 per cent increase from 2009, +20.09 per cent if compared with 2007.[8]

The second-hand market and the development of the offer through servitization[9]

2009 saw a quick development of the market for second-hand forklifts, which suddenly gained relevance due to the crisis and a shift in the needs of operators, market makers and the logistics chains connected to them.

Until that time, the distribution organizations did not really assess the viability of setting up a structure that could properly support this new market, helping them to exploit the opportunities that arose.

Beginning from 2009, an increasing number of managers of distribution companies started taking into consideration the development of a structured organization for the second-hand market, increasingly fuelled by the transformation of the original main business of selling new units: traditional sales were now beginning to be integrated with a new set of extra services. Companies adopted new solutions for short- and long-term rental agreements, launching new products, services and business solutions: servitization started being an important factor in the industry. Shifting from sales to rentals activated a series of evolutionary passages in the supply chain, which not only needed new abilities, skills and knowledge, but it also activated a process of business regeneration after the first rental time cycle (five years): now there was a new set of units to work on and re-launch in the circular management of used and guaranteed forklifts.

The importance of the role played by the management of used trucks is highlighted by the growth of the management of company fleets, which require substantial resources, both in investments and organization, and increasingly need a direct intervention of the commercial organizations of forklift retailers. The consequence of this new business approach is the creation of units with direct sales organizations, with branches in the territory that can help and support the dealerships, paying them commission for the assistance given while signing new deals with the customers, shouldering the financial side of rental projects and programmes (which the dealers are not able to control directly due to their financial limitations), leaving the dealers to handle emergency and technical support and maintenance of the rented units.

Notes

1 As shown by Figures A.3a/b, A.4a/b, A.5a/b and A.6a/b, included in Appendix A, which show the sales in the Italian market during the 2007–16 period.
2 Figures taken from Appendix A, Figures A.3a and A.3b.
3 See Appendix A, Figures A.4a and A.4b.
4 See Appendix A, Figures A.5a/b and A.6a/b.
5 See Appendix A, Figures A.3a and A.3b.
6 See Appendix A, Figures A.5a and A.5b.
7 See Appendix A, Figures A.4a and A.4b.
8 See Appendix A, Figures A.6a and A6b.
9 Intended as the expansion of offer, value and business with the addition and integration of services and solutions.

23
Speech by Leonardo Salcerini

Toyota Material Handling Italia, represented by Dr Leonardo Salcerini, faces the convention of 9–11 October 2009 with two sales networks to merge, while experiencing a devastating economic crisis that has halved sales in the space of only two years.

This is the introduction by Leonardo Salcerini:

> Dear dealers,
>
> We all are fully aware of the current situation; we are facing a serious and complex time in our corporate history. However, if we could agree on a strategy to implement together, I am positive we could mitigate our damages and lay the foundations for growth; with the right risk management strategies, we can begin our recovery and ensure our future profits.

The introduction by Leonardo Salcerini is rich and detailed, including topics such as:

- the strategies to implement for the merger of the sales networks;
- the tactics and strategies for branding;
- the multiple and complex policies regarding marketing, distribution and sales, explaining the values of the Toyota Way which will drive the operational and managerial procedures of the new commercial organization, while being at the core of the future growth of the company.

Foreword

The situation of the Italian market for forklifts at the eve of the convention

Before the convention was due to begin, the **Italian market scenario** for Toyota had a couple of main issues:

a sales were in freefall, crashing from 46,076 units sold in 2007 to just 24,427 in 2009;

b dealing with the merger of two separate indirect sales networks, 85 dealerships that until yesterday were part of two distinct businesses fiercely competing with each other.

The sales networks were in turmoil, concerned about the growing complexity of the market and the upcoming reorganization, which added to the other issues they had to face:

- the credit crunch that froze financing;
- the crash of market volumes;
- the transformation of the market, which was migrating from straight sales to rentals;
- the increase in competition from other brands and among the same dealers of the company's networks, which sold slightly different products in the same areas.

The market share crash: the need to grow and recover such share

BT CESAB and Toyota together had a combined 22 per cent market share in 2007; in 2009, it had fallen to a meagre 17.5 per cent.

Both brands had an urgent need: to grow and recover the market share they had lost!

Pressing issues and dilemmas related to the unification of indirect networks

The management of Toyota Material Handling Italia (TMHIT) was facing some serious issues, the solution of which was crucial to determine:

1. the success, or failure, of the multi-brand policy of the company;
2. (in addition to) the success of the entire strategy of the company;
3. leadership and reputation of the business;
4. business performance and continuity: making the wrong choices for the sales networks would have caused the defection of dealerships, which in turn would have badly affected performance, revenue and the actual margins on sales.

The failure to find the right choices would have led to a tipping point: the situation would have affected future turnover, jeopardizing growth and questioning the capabilities and the leadership skills of the management itself.

Such a complex situation of the markets created a series of problems that required a complex and system-wide intervention by the top management.

Here we list the main issues and the questions that still needed to be answered.

Main doubts: problems to solve and 13 open matters

1. Which were the relevant and immediate choices to be made immediately regarding the sales networks?
2. What needed to be done to avoid defections from the sales networks?
3. Was it still viable to maintain two separate sales networks with a multi-brand policy, or was it better to merge the networks?
4. Which were the risks to avoid, the issues to prioritize and the courses to take?
5. How should the networks be managed during this crisis?
6. How should the territory managers be organized, with their different backgrounds and managerial culture?
7. How can the financial difficulties of the dealerships be overcome and market share expanded at the same time?
8. How can the transition of politics, competences and capabilities of the dealerships, which had to deal with a major shift from sales to rental, be supported? Moreover, how can the second-hand market that would derive from the expansion of rental agreements be organized?
9. How can growth be achieved while avoiding clashes between dealers belonging to different sales networks?

10 How can aggressive, competitive multi-brand policies with a range of slightly different products that are sold in the same areas be implemented?

11 How can we avoid dealerships cannibalizing each other to exploit business opportunities?

12 What do we do to grow and recover the market share that was lost?

13 How can we protect and balance the brand policies and at the same time ensure operational continuity to the dealers located in the same areas, while they are being affected by both the credit crunch and the devastating sales slump?

The contextual solution of all open issues required setting up a number of strategies and an innovative, daring and winning cooperative approach, one that could minimize risks and ensure performance, harmony and winning synergies.

The new cooperative approach followed the steps and values of the Toyota Way, in a process of contextualization and cultural, strategic and operational declination, up to its executive implementation.

The Toyota Way vision of values and systemic perspective

As is widely known, the Toyota Way is a substantial element of corporate culture; it transpires in the modus operandi of those who follow it, which eventually is seen as a natural declination and is consequently shared by everyone, allowing the creation of initiatives and tools that continuously improve all of the company activities.

In order to see it as an element that influences all activities, we can see the Toyota Way as a cultural homogeneity made and driven by two main subsystems which are strictly intertwined:

a the value subsystem of **employee development**;

b the value subsystem of **continuous improvement**.

The value subsystem of employee development is itself composed of the core values of **respect** and **teamwork**. Continuous development is again split into its constituted value dimensions:

1 accepting challenges and undertaking challenging initiatives;

2 delving into the origins of the issue, stimulating people to understand the situation properly and solve it at its root (*Genchi Genbutsu*);

3 stimulating the flow of value generated by constantly improving processes and actions (*Kaizen*).

Differential interpretation and the addictive qualification process of the Toyota Way's value systems

The application of the Toyota Way is performed through a common effort, finalized to reinterpret Toyota's corporate culture in each organization. The Toyota Way cannot work only by applying its principles without taking into consideration the company subject to such transformation: it must be moulded and harmonized according to the kind of organization concerned.

Here follows a summary of Salcerini's speech at the convention.

The new unification project of the two sales networks (BT CESAB and Toyota) has some unique features, strictly related to the principles of Lean organization and of the values of the Toyota Way, which are summarized in the four cornerstones outlined below. Such cornerstones define and direct business choices, market strategies, the guidelines of the new TMHIT distribution policy and the path to follow by the new, integrated, indirect sales network:

1. The Toyota Way and Lean organization: differential, additional and multi-focal implementation.
2. One channel, three brands.
3. GAP: provincial harmonized management.
4. Real transformation of the commercial policy and of business and sales approaches.

The Toyota Way and Lean organization: differential, additional and 'multi-focal' implementation

The principles of Lean organization and the values of the Toyota Way were first decoded by us, adapting them for our Italian business culture, and contextualizing them on the basis of the dynamics and complexities of the Italian forklift market, which includes logistics activities by industrial and commercial customers. Finally, we integrated and implemented Toyota's philosophy and culture values, which are the founding blocks of the Toyota Way and define the development of employees (generated by the centrality given to respect and teamwork) and the idea of continuous development (itself composed of the values that support the *Genchi Genbutsu*: going to the core of things).

These activities are what created the new added value in TMHIT when it applied the guidelines derived from the Toyota Way.

The added value of TMHIT management in processing the principles of Lean organization and the Toyota Way

Through the processing of knowledge, of the Toyota Way values and of the principles of Lean organization, TMHIT management proposes new ideas that, in turn, contribute to add to the corporate vision, identifying new landscapes that can create alternative strategies. In other words, a new and unique interpretation of such values and principles will support the creation of a new corporate mission, without corrupting the previous constitutive factors.[1]

The added value and the differential implementation formulated by TMHIT and Toyota Academy reside in the processes that we called D.> C.> A.> I.+, an acrostic that summarizes and introduces the nuclei of activities that generate operational differentiations and qualifications, giving the Toyota Way values new strategic and operative declinations, capable of enriching the traditional values and driving transformation and development of TMHIT's commercial strategies.

The initiatives that were implemented in the context of the new D.> C.> A.> I.+ processes concern:

- Decodifying the values of the Toyota Way, adapting them to the Italian corporate culture;
- Contextualizing the same values of the Toyota Way for the economic landscape in Italy;
- applying the Applicative dimension of the Toyota Way values to the forklift business in Italy;
- Integrating and Implementing the values of the Toyota Way into the new strategies and into distribution and commercial policies, to develop the processes needed to unify TMHIT sales networks and optimize the capability of exploiting business opportunities at different territorial levels (national, regional, local).

In particular, the 'multi-focused' operative declination deals with:

- people's orientation and centrality;
- long-term vision and actions;
- the challenge in the long term: qualified collaboration among all the dealers, in particular within the new sales organization set up by TMHIT, specifically in its new integrated sales network, in a close partnership with all the other elements of TMHIT, dealerships and customers;
- focusing on the respect that defines relationships among local dealers, between TMHIT and dealers and customer networks, which implies new principles in the management of the sales network that will be characterized by transparency, fairness, correctness and equal treatment of all dealerships. Such values shall prevail in all relationships among local dealers.

Furthermore, new definitions are given to:

- new tactical and strategic approaches for the commercial processes of TMHIT, with the development of a new Lean distribution organization: a new network which is 'peaceful' and harmonious, born from integration and a new collaboration process, an organization of the dealers' networks that is structured and regulated, capable of satisfying the ever changing and diverse requests by the customers, reflecting a demand which is increasingly specific and diversified;
- a new, far-sighted vision for the commercial organization, capable of offering an overview in the long term and an efficient local structure;
- a specific focus on the evolution of cooperative distribution strategies among dealerships, competitors and promoters of brands that were competing against each other up to yesterday, so that they can successfully implement a constructive integration;
- a new distribution organization where the single dealerships switch from a single brand to the promotion of two separate manufacturers (BT CESAB and Toyota).

One channel, three brands

All over Europe, Toyota Material Handling is following a commercial strategy focused on 'one channel, two brands'.

Italy has a more complex situation that can bring further opportunities, because TMHIT has to manage three brands with one single channel.

We could have chosen three different, alternative strategies to merge the sales networks:

a select the best dealerships for each area, discarding the others;

b apply a joint effort with production and marketing and apply a marked differentiation between the products manufactured by Toyota and by CESAB;

c follow more closely the principles of Lean production and the Toyota Way, making decisions based on people's leadership, on the trust guaranteed and generated by top management towards operators and employees and by those who work in the sales networks.

TMHIT decided to embrace the third option, on the basis of the mutual respect and on the relationships that management built with dealers and distributors for over 10 years, through a continuous and supportive presence. Such favourable conditions reassured TMHIT and gave it the strength to adopt a more revolutionary approach.

The option chosen was by far the most complicated to implement: it relied on a successful cooperation between Toyota and the dealerships, with the knowledge that most of the evolution and performance of all the companies of the group would be deeply affected by how well this cooperation worked. This choice immediately discarded the possibility of terminating the relationship between the company and some of the dealers. Maintaining the current setup and merging the sales networks can help to exploit the business opportunities that could arise at a local level; maintaining the different elements of the dealers' distribution organizations would help to carry out the transformation without interruptions, supporting the growth in performance by focusing even further on the customers.

The integration of the sales networks shall be able to reduce any duplications and waste; it will allow us to improve our coverage of the areas, offering to dealers further options for their customers, widening the range of brands, products, services and solutions to sell on the market. Moreover, this choice will make the dealerships more attractive, giving them the chance to sell both brands and equipping them with new tools to recover sales and achieve further growth. The optimization of the commercial network and the effective multiplication of the products on offer will give the dealers more chances to satisfy the customers' needs, which in turn will help get more sales and business.

The new distribution organization, while it will mainly work alongside the customers (the same small to medium clients), will constitute a new entity integrating all the identities, cultures and organizations previously operating.

This strategy will allow us to:

- improve our presence in the territory, boosting the chances to recognize and exploit business opportunities;
- maximize the efficiency of the organization while managing the business opportunities that arise on a local level;
- optimize the split management of the brands, in an effort to qualify and differentiate their offers;
- develop and increase sales and productivity of the dealerships in the short, medium and long term.

GAP: Gestione Armonica Provinciale (provincial harmonized management)

The development of a new local distribution policy, implying the merger of the dealership networks, needs the **respect** of the activities undertaken by each dealer; first of all, it requires mapping and analysing the sales for each area covered by the dealerships.

In order to focus on the analysis concerning the development of distribution, we need to assess the customers served by each location, setting up an organic management of the customers of each dealership at a provincial level. This would avoid dealers chasing the same customers, which would in turn cause debilitating internal wars to acquire them. The management focus will ultimately shift from the product (and from the traditional business trade marketing) to an efficient management of the logistics chain and of the final users.

The offer will be more specialized, moving towards differentiation, with integrated solutions improved by several new financial and logistics services.

The GAP and the new role of the regional manager

The existence of two separate sales networks implied the presence of two different area managers in each area where BT CESAB and Toyota were operating simultaneously.

With the integration of the networks, the new territorial management requires the addition of a single regional manager, who will have the following roles:

- equilibrator, coordinator, sole 'governor' and impartial arbiter, in charge of reducing interferences and overlapping between dealerships;
- supervisor and arbitrator for any disputes that might arise between dealerships;
- sole responsibility for the relationships with the key accounts managed directly from the commercial management in Bologna, Italy;
- inspector in charge of ensuring the respect of priorities and activities of each dealership located in the reference area.

It is agreed that this structure needs a responsible person who is in charge of supervising the respect of the rules of commercial behaviour, ensuring harmony among dealers and all the organizations involved in the logistics chain business processes operating in the reference area.

Tangible transformation of commercial policies and business and sales approaches

This new approach by Toyota Material Handling Italia contemplates a tangible transformation of the commercial policy and of the business and sales models, which will develop on two main levels:

a the first will be the introduction of a new, comprehensive range of offerings;

b the second will involve a new form of flexibility and several options to choose in terms of commercial policy and 'go to customer' configurations.

The first level of transformation of TMHIT commercial policy: the introduction of a new, comprehensive range of offerings

The first step in the transformation of our commercial policy implies the introduction of a new package of products and services that will ensure more business opportunities and bring a tangible increase in revenue for the dealerships.

In addition to the traditional sale of a package of products and services that will define the business and increase revenue and profits for dealerships, TMHIT is suggesting:

- a new fleet of forklifts for short-term rental (new offer/service);
- new forms and formulas to finance the offerings;
- the development and expansion of the product portfolio and the creation of the **'one-stop shopping'** formula: thanks to central purchasing agreements with a cluster of suppliers of accessories (shelves, batteries, tyres, etc), TMHIT will be the 'central purchasing unit' that will contribute to the transformation of the company's commercial offer, which will not be limited to the forklifts but will include the whole range of products and solutions on offer: their extension will offer more business opportunities to the dealers and give an appropriate answer to the new, increasing needs of the customers.

In brief, TMHIT will complement the traditional activity of forklift sales with new lines of products and services, offering to the dealers' network:

- new consulting and specialized training services;
- financial support and services;
- a short-term forklift rental service for the dealers and their customers, capable of meeting the needs of the dealerships and limiting their financial exposure without missing business opportunities;
- a set of agreements with other suppliers of accessories and ancillary products, to integrate and differentiate the traditional offer, making it more attractive for the customers thanks to more flexible and competitive prices, with the goal to create further benefits and value for the dealerships.

The second level of transformation of TMHIT commercial policy

A further, significant level of transformation of the commercial policy will be the involvement of TMHIT in sales, in support of the dealers; the aim is to exploit all possible business opportunities, sustaining the increase in turnover, maintaining existing customers and attracting new ones. The commercial transformation implies more freedom of choice for the dealerships, which will have the opportunity to play a role of sales agencies in certain situations, reducing their financial needs and reducing their commercial action to a report regarding the business opportunity.

In the event that the business opportunity is extremely relevant for the dealership, but it entails a financial burden that might be too heavy to support, there are two possible options:

- act as a traditional dealership: buy the trucks from TMHIT and subsequently sell or rent them to the customer;
- act as agent: report the opportunity to TMHIT.

The second option can simplify and reduce the level of risk for the dealer, while avoiding the loss of a number of opportunities both in the short and long term.

Acting as a sales agency will allow the dealership to limit risk and financial exposure, while still seizing the business opportunity and acquiring the customer.

This option will furthermore allow the dealer to gain additional profits from the commissions due as agency; moreover, it will have the opportunity to supply the maintenance and technical support services that will eventually be required by the customer.

In such cases, invoicing will be done directly by TMHIT. The dealership will be rewarded a commission on the transactions (sale or rental).

Conclusions

In brief, what happens with the shift of TMHIT's commercial policy?

a More business opportunities for dealerships: a new offer with a wider range of products, more solutions and additional services.

b A simpler, unified management: one single person to refer to, someone who is in charge of supervising, deciding and coordinating all the commercial activities in the territory.

c A new organization, which preserves the precious customer base of the dealers and sets new, clear rules for everyone to follow.

d A single pricing policy for everyone, with the same discounts and conditions.

The transformation of TMHIT's commercial policy will therefore guarantee an unprecedented level of flexibility to its dealers. They will have the possibility to choose whether to be involved in the business as sellers or intermediaries, according to the size of the opportunity and their needs.

The role of intermediary will require new key account functions and activities, which the regional manager will follow to coordinate and manage all the sale and rental agreements stipulated directly with the end user.

These new transformation and development steps of TMHIT's commercial policy will be reinforced with new rules of behaviour and organization that will coordinate all the local initiatives and activities of the dealers.

It will be crucial for everyone to respect new guidelines that will involve new techniques to approach the customer and the market in general. It will also be necessary to respect the activities undertaken by each dealership, alongside the development of new commercial and consulting skills. Respect and cooperation will ensure the harmonious and continuous growth of profits.[2]

Notes

1 As stated by Aristotle in his *De generazione et corruptione* (*On Generation and Corruption*), there is corruption only if the offence is transformed so much, and so deeply, that its nature is no longer known. Something is corrupted when there is a transformation of its constitutive factors and permanent elements. The term 'corruption' here is intended as a radical transformation.

2 In the meantime the weather had changed: the morning's thick clouds had disappeared, and the sun was warming up the air. The hotel's maître d' had appeared in the congress room, announcing the lunch break. The worried faces and the concerned looks that were aplenty at the beginning of the event had given way to smiles and signs of relief. The dealers had shown their faith in the passionate and inspiring words they just heard from Leonardo Salcerini. A long, spontaneous applause broke the silence and filled the room. 'Work will resume after lunch at 2pm', were the last words that echoed in the pleasant room of Riccione's Hotel Atlantic. The first part of TMHIT's convention was over.

24
The new ACCESS model

The interpretation of the Toyota Way and its contextualization in a medium-sized Italian company

We felt that it was important to apply the precepts of the Toyota Way to manage the creation of this new, untested, fragile distribution system.

It was fundamental for us to face this complex issue with the most appropriate approach; there was a significant risk of conflicts between dealerships that were, after all, fierce competitors right until the convention took place. We needed a strategy that was detailed and carefully planned. We had to set up a unique tactical and strategic reference model which could unite and unify.

The teams of Toyota Material Handling Italia & Academy, after a long analysis and research that involved a large number of dealerships, alongside detailed studies on a sample of final customers belonging to the industrial and large retail sectors and an in-depth study of the competitors, had prepared a new approach model and a new people and distribution management system.

The new ACCESS model for a 'coop-active' approach

This new model, called 'ACCESS', is based on the values of the Toyota Way and on the contextualization and application of its principles, in order to:

1. inform and characterize the new distribution strategies;
2. govern the complexity of the system;
3. simplify problems;
4. guide dealers, employees and business choices;

5 drive innovation;

6 reduce the risks deriving from business choices;

7 answer questions and solve doubts;

8 ensure the leaderships of company and dealers;

9 guarantee the expected performance and the effectiveness of the solutions adopted.

Figure 24.1 The Toyota Way: differential declination of Toyota values, their contextualization and the new 'coop-active' approach to marketing and sales: the ACCESS model

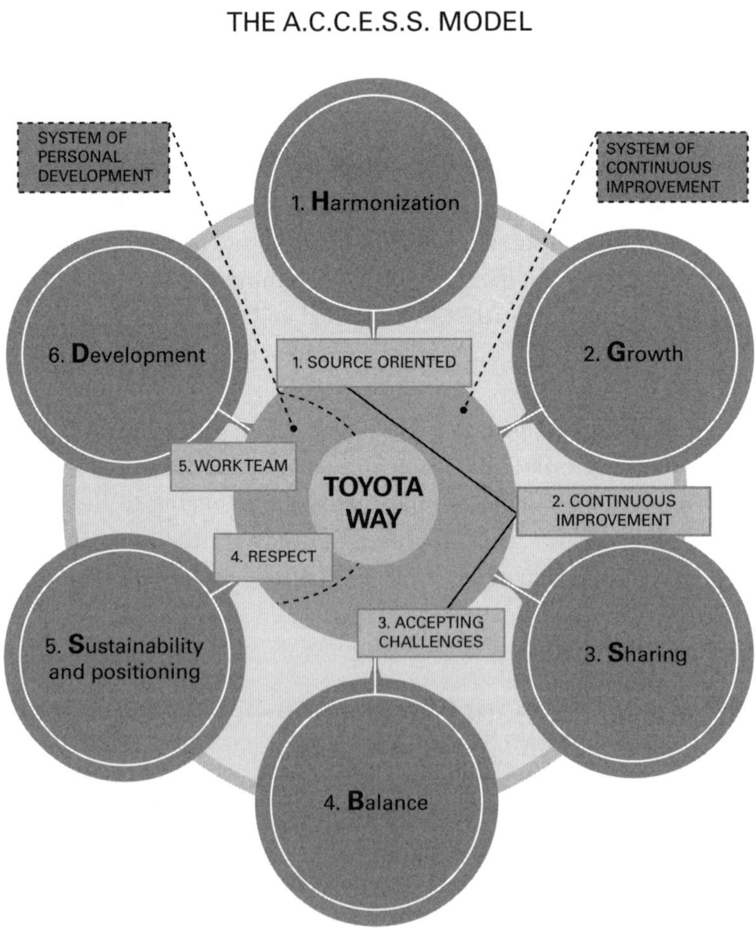

This model, designed to simplify the process while generating value for all the stakeholders, could be summarized with the ACCESS acrostic. Its main concept is to define and develop organizational policies and processes inspired by Toyota values, namely respect, teamwork, finding the source of the problem, accepting the new challenges of the market, and improving continuously.

The application of the new model of collaborative approach to the market could be crucial to set up the new strategies and to define the strategy and organization that would be the reference behind the managerial choices taken by Toyota Material Handling Italia.

Part Three
Appendices

APPENDIX A
Data

Figure A.1 shows the map with the location of BT CESAB (dark dots) and Toyota (light dots) dealerships in Italy on 8 October 2009.

Figure A.1 Geographic distribution of BT CESAB and Toyota sales networks in Italy on 8 October 2009

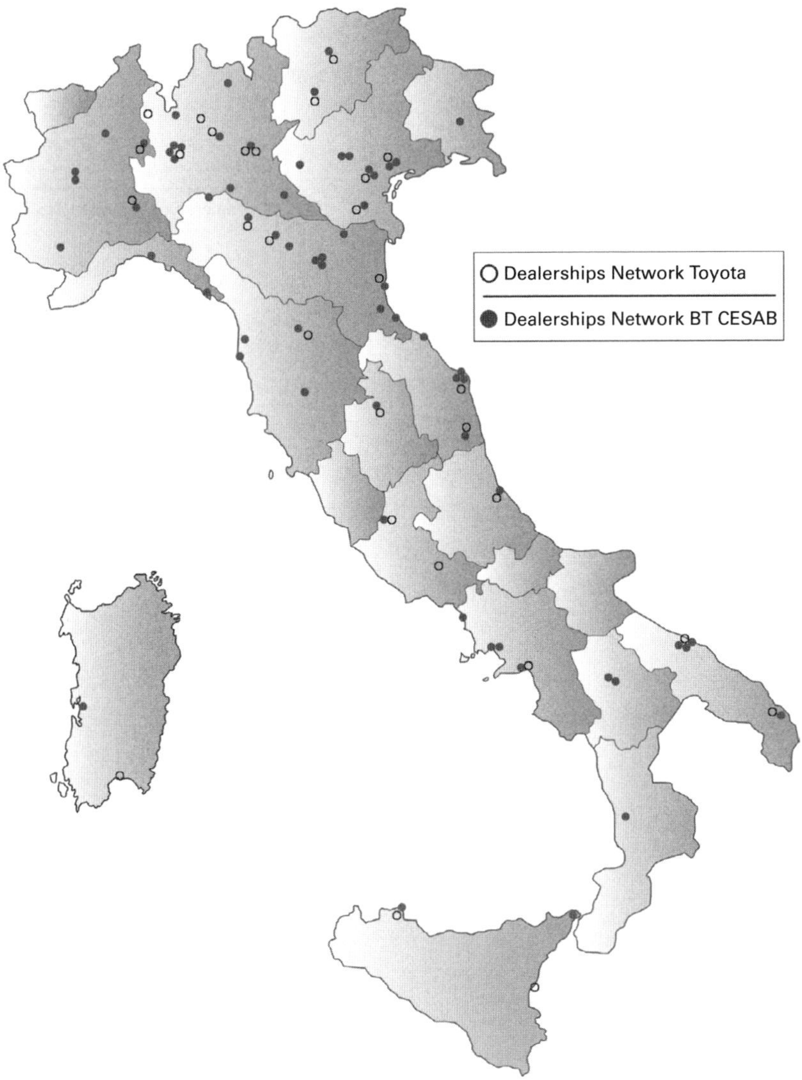

Figure A.2 shows the map with the location of BT CESAB and Toyota dealerships in Lombardy on 8 October 2009.

Figure A.2 Geographic distribution of BT CESAB and Toyota dealerships in Lombardy on 8 October 2009

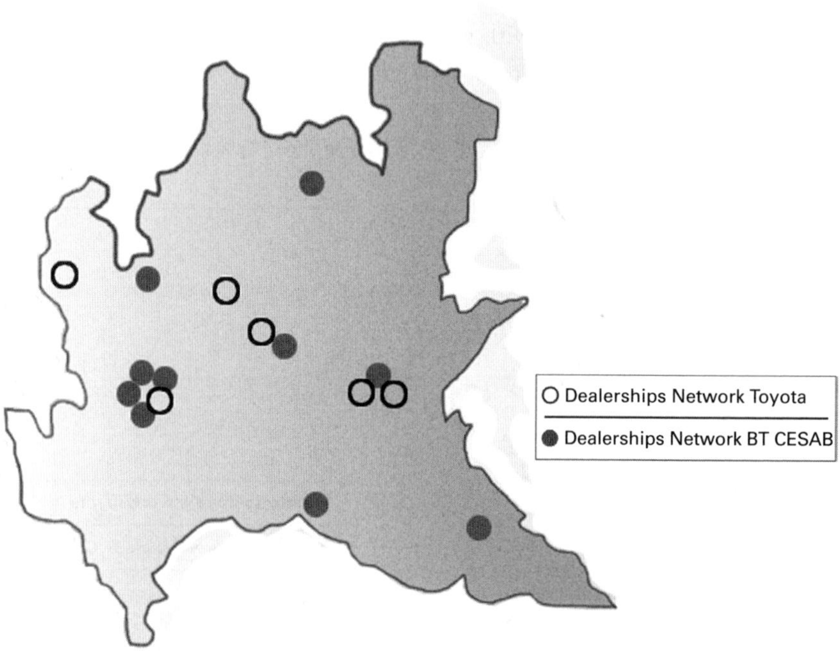

Figures A.3a and A.3b show the figures of the entire Italian forklift market (2007–09).

Figure A.3a The market for forklifts in Italy (2007–09)

TOTAL ITALIAN MARKET FOR FRONTAL FORKLIFTS 2007–2008–2009 (*)

N.	DESCRIPTION	2007	2008	VAR% 2007/2008	2009	VAR% 2007/2009
01	FRONTAL ELECTRIC TRUCKS	16.525	13.450	–18,60%	6.540	–60,4%
02	DIESEL AND GAS POWERED FRONTAL TRUCKS	6.148	5.458	–11,2%	2.643	–57%
03	TOTAL FRONTAL TRUCKS	22.673	18.908	–16,6%	9.183	–59,5%

* Values expressed in units

NOTES World Industrial Trucks Statistics data

Figure A.3b Total market for forklifts in Italy (2007–09)

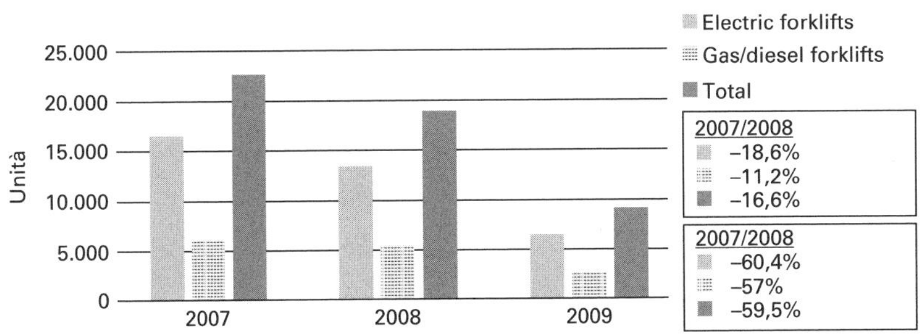

The forklifts taken into consideration in Figures A.4a and A.4b are mainly destined for distributors, large retailers, logistics services and warehouses.

Figure A.4a The market for warehouse forklifts in Italy (2007–09)

APPENDIX N. 04

TOTAL ITALIAN MARKET FOR WAREHOUSE FORKLIFTS 2007–2008–2009 (*)

N.	DESCRIPTION	2007	2008	VAR% 2007/2008	2009	VAR% 2007/2009
01	WAREHOUSE, LARGE RETAIL AND WAREHOUSE TRUCKS	23.403	20.985	–10,03%	15.244	–34,09%

* Values expressed in units

NOTES World Industrial Trucks Statistics data

Figure A.4b The Italian market for warehouse forklifts (2007–09)

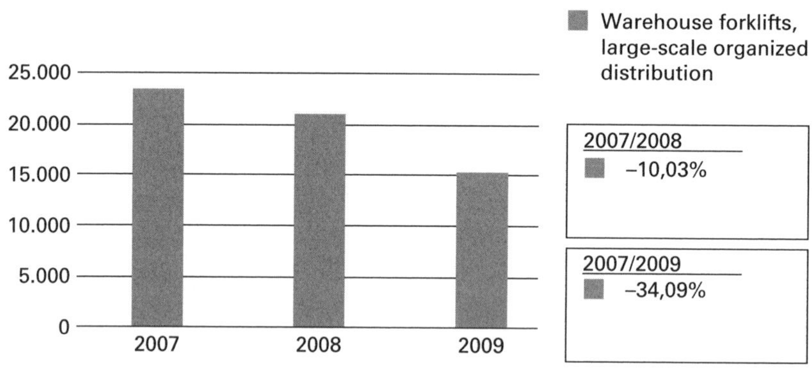

Figures A.5a and A.5b show the total market for forklifts in Italy in the long term, 2007–16.

Figure A.5a Total market for forklifts in Italy (2007–16)

APPENDIX N. 05
THE ITALIAN TOTAL MARKET OF FRONT TROLLEYS -
EXTENDED ANALYSIS IN THE LONG PERIOD 2007–2008–2009 (*)

N.	DESCRIPTION	2007	2009	2016	VAR% 2007/2016	VAR% 2009/2016
01	ELECTRIC FRONTAL TRUCKS	16.525	6.540	13.479	**–18,44%**	**+ 106,10%**
02	DIESEL AND GAS POWERED FRONTAL TRUCKS	6.148	2.643	3.509	**–42,93%**	**+ 32,77%**
03	TOTAL FRONTAL TRUCKS	22.673	9.183	16.988	**–20,07%**	**+ 84,99%**

* Values expressed in units.
NOTES World Industrial Trucks Statistics data

Figure A.5b Total market for forklifts in Italy (2007–16)

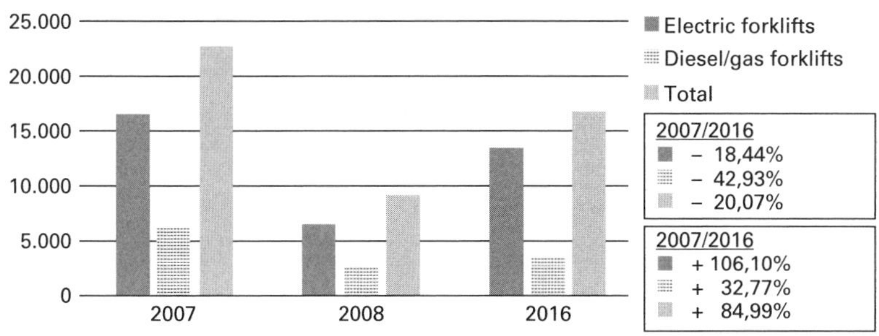

Figures A.6a and A.6b show the total market for warehouse forklifts in Italy in the long term 2007–16.

Figure A.6a Total market for warehouse forklifts in Italy (2007–16)

APPENDIX N. 06
THE ITALIAN TOTAL MARKET OF FRONT TROLLEYS -
EXTENDED ANALYSIS IN THE LONG PERIOD 2007–2009-2016 (*)

N.	DESCRIPTION	2007	2009	2016	VAR% 2007/2016	VAR% 2009/2016
01	WAREHOUSE, LARGE RETAIL AND LOGISTICS TRUCKS	23.403	15.244	28.296	**+20,9%**	**+85,62%**

* Values expressed in units
NOTES World Industrial Trucks Statistics data

Figure A.6b Total market for warehouse forklifts in Italy (2007–16)

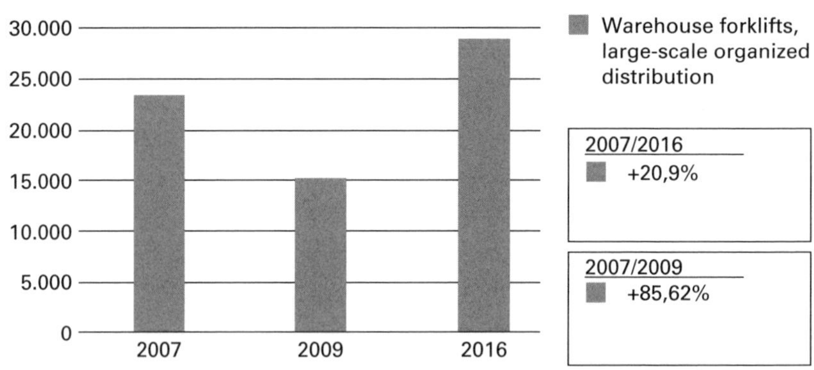

Figures A.7a and A.7b show the analysis of market share for the period 2009–15.

Figure A.7a Market share analysis for the 2009–15 period

APPENDIX N. 07
MARKET SHARE ANALYSIS FOR THE 2009–2015 PERIOD

YEAR	2009	2010	2011	2012	2013	2014	2015
MARKET SHARE	17,5%	17,3%	18,6%	20,95%	22,59%	23,19%	23,44%
WAREHOUSE, LARGE RETAIL AND LOGISTICS TRUCKS	17,4%	17,3%	18,9%	21,26%	23,76%	23,20%	23,22%
ELECTRIC FRONTAL TRUCKS	20,7%	19,6%	21,4%	23,46%	23,81%	26,17%	31,96%
DIESEL AND GAS POWERED FRONTAL TRUCKS	10,4%	10,8%	8,7%	9,82%	9,23%	12,15%	11,88%

(*) WITS DATA

NOTES World Industrial Trucks Statistics data

Figure A.7b Market share analysis for the 2009–15 period

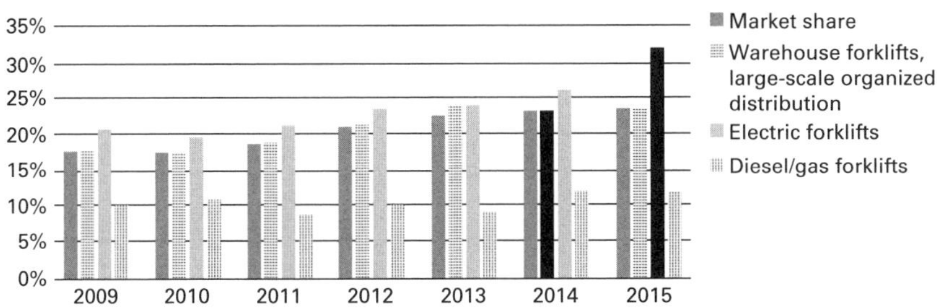

APPENDIX B

Reader's guide to the universe of Toyota Material Handling in Italy

by Leonardo Salcerini and Danilo Verga

The main goal of this appendix is to offer more information to the readers, so they can better understand the methods and best practices adopted by Toyota Material Handling in Italy that allowed the company to achieve successful results. This appendix can also be implemented as a reference tool for the application of some of the methods and corporate transformation practices we introduced, which can be customized and adopted by a wide range of companies operating in several sectors.

This appendix also aims to:

- suggest a number of considerations that can help detect key elements that are capable of giving useful perspectives for corporate growth and development;
- offer a key to interpret the complex structure that eventually brought continuous success and excellent results, with the aid of diagnostic and operative tools that can support the implementation of change and transformation processes;
- contribute to detect the main issues to and create new perspectives for the growth and development of strategic guidelines;
- facilitate the business assessment to identify new paths, options and corporate transformation opportunities.

Business transformation and innovation process: continuous success and the concept of the new cooperative management 4.0

The synthesis of the performance and the results obtained with the complex and structured continuous innovation transformative process (shortened to TIC, Trasformazione Innovativa Continua, or Continuous Innovative Transformation) highlights the continuous success obtained with the intervention of Italy's Toyota Material Handling managers in the 2000–16 period.

This success is testified by the length of time taken into consideration (over 15 years), and by the consistent, ongoing results obtained by the managers of Toyota MH.

This success, which we can define as 'continuous' for its entity and duration, was achieved with the appropriate adoption of Toyota's managerial platform (built with and inspired by the principles of Lean leadership, by TPS management processes and by the Toyota Way values), facing the challenges that arose and making specific, game-changing decisions.

The continuous corporate success, derived by the excellent managerial results that were obtained in production and sales, was achieved thanks to the application by management of a winning managerial formula, whose characteristics are defined in the so-called 'new cooperative management 4.0'.[1]

A summary of the main indicators of continuous success of Toyota Material Handling in Italy

The following is a synthetic introduction and summary of some of the main areas and continuous success indicators:

1. occupation, and the continuous success of personnel employment and training;
2. continuous production and organization of operational and corporate processes;
3. continuous optimization of operation processes: warehouse turnover;
4. continuous success achieved with the definition of suppliers' relations and their inclusion in the corporate transformation success;

5 continuous success of the commercial innovation and transformation processes: the development of aggregate revenue;

6 continuous success of the commercial innovation and transformation processes: the acquisition of market share;

7 development and extension of the continuous success in the corporate transformation to the indirect sales networks, to external participants and to the logistics operator chains (definition, support, reorganization, fidelization, innovation and extension of business and partnership potentials and opportunities);

8 development and extension of the continuous and transformative success outside the company, towards the business community, with the initiatives of Toyota Academy;

9 development and extension of the continuous and transformative success outside the company: the transformation of the network of relationships in the territory and with other stakeholders (schools, universities, etc).

1. Occupation and the continuous success of personnel employment and training

The employment rate is by far the most important factor for national and local economies: its increase, in particular among young people, is one of the indicators of social success of a company, because it represents the contribution of the business to the creation of wealth in the territory. The employment rate was one of the challenges that Toyota Material Handling successfully faced in its transformation: the corporate changes that were applied limited the use of welfare support and actually created new jobs, in an environment of constant professional improvement. The crisis that began in 2008 and affected all economic sectors was faced and overcome by the company: the number of people employed by Toyota Material Handling in Italy has risen from 344 in 2000 to 1,353 in 2016, with a 393 per cent increase.

The Italian branch of Toyota Material Handling is Toyota's European sales company with the highest percentage of women in the workforce.

Furthermore, in spite of the merger and optimization of the sales networks, there has been no reduction of dealerships, even with the consequences of a terrible crisis and of the subsequent selection process that reduced the number of enterprises in the territory.

2. *Continuous production and organization of operational and corporate processes*

As we discussed in the first part of this book, one of the most notable revolutions created by this business transformation has deeply affected the corporate culture and the system of competences, transforming operative processes and local production into a highly efficient environment with strongly innovative standards and international excellence.

Such transformation was possible with the application of Toyota's production methods, widely known and appreciated worldwide, integrating them with the principles and the values that identify the entire value-generating platform that is typical of Toyota (Lean leadership, Toyota Production System and the Toyota Way), with the added value of the unique managerial contribution by the Italian management of Toyota Material Handling.

This shift in corporate culture has deeply modified the system of continuous learning for the people inside the company, and partly for those outside. A medium-sized local manufacturing enterprise went through a radical transformation that made it become a modern organization and an international market leader, capable of learning and improving continuously (a qualified advanced deep learning organization), a global trend-setter for management techniques and a reference point for all manufacturing and commercial businesses, be them local or global.

The same transformation of corporate culture had a massive impact on the system of knowledge, competence and skills of the entire workforce of the Italian organizations of Toyota Material Handling, on the company's operational conduct and on people's behaviour. It has also changed the organization of work, and affected internal and external communications with the introduction of new systems for continuous interaction, exchange and feedback, both within the plant's organization and with the other departments and sectors. Such activity has spread outside the company, extending the new approach and methods to the customers and suppliers' organizations.

In particular, a specific effort was made to select and qualify our relationship with the suppliers, with the aim of producing efficiency and continuous innovation.

Our interaction with the organizations and the employees of the suppliers made it possible to transfer the 'pillars' of the new corporate culture of Toyota Material Handling to the employees of the supplier's organization.

Such cultural contaminations and transfers of knowledge, aimed at defining relations and organizations in order to achieve optimization and continuous efficiency of the supplier's offer, have eventually affected the organization of the supplier itself, creating a flow of new value and setting up a win–win movement towards continuous improvements.

Finally, the introduction and development of a new attitude, motivated by change, by continuous improvement, by sharing and collaborating, have brought high productivity levels, a new innovative, collaborative environment, alongside the value-creating awareness derived by a strong sense of belonging to the company.

3. Continuous optimization of operation processes: warehouse turnover

The transformation of the company and the search for continuous efficiency, innovation and optimization have brought excellent results, generating value and widespread efficiency in all the company sectors. A good example is given by the success of business transformation processes, summarized by the evolution of the warehouse turnover index, which increased from 2 in 2000 to 26 times in 2016. Such positive evolution has reduced the total amount of working capital, aside from other benefits and cost savings in terms of reduction of the required area for storage, creating new levels of efficiency and productivity.

4. Continuous success achieved with the definition of suppliers' relations and their inclusion in the corporate transformation success

As previously noted, our company transformation has produced excellent results with our relationships with suppliers: a process of continuous optimization, capable of creating value and benefits for both participants of the supply chain, namely the supplier organization and the customer/partner company.

The activity of business transformation has expanded, involving continuous qualification and optimization of suppliers, always searching for increasingly higher efficiency levels. Toyota Material Handling Manufacturing Italy has entered into a constructive dialogue and interaction with its suppliers, applying a systemic strategy of selection, relationship and operational study of the suppliers themselves. The encouraging results we obtained with this transformation guided the selection of suppliers, which were reduced to 180 from an initial number of 350. Such a selection process has increased the contribution of the suppliers to business activities: their involvement has grown from €60 million to €300 million in 2016.

5. Continuous success of the commercial innovation and transformation processes: the development of aggregate revenue

Our steps towards continuous innovation, the evolution and qualification of managerial processes, a constant, constructive communication between the commercial organization, the factory and direct customers, have created an environment based on trust by dealers and customers regarding the management of the commercial organization; this positive aspect drove a continuous virtuous transformation that supported an ongoing increase in turnover.

During the long period subject to our analysis (2000–16), a number of innovations, revolutions and changes have transformed the business of forklifts in Italy and Europe:

- stronger competition by other manufacturers;
- significant evolutions and transformations of the business models, which shifted the focus from the sale to the rental of forklifts;
- a drastic structural change of the forklifts market, partly caused by the 2008 demand crisis;
- a massive rise in customization requests by the customers, who increasingly look for specific solutions;
- an indispensable transformation of sales processes and consequently of the skills and commercial abilities of the individuals involved in the commercial area;
- positive returns from the distribution choices implemented by TMHIT management, complying with the new strategic commercial approach: the company turnover has constantly improved, rewarding the decisions that were taken to face the challenges posed by the market.

In mere quantitative terms, Toyota Material Handling Italia has posted an aggregate turnover that grew from €200 million in 2000 to €730 million in 2016, with an average increase of 365 per cent.

6. Continuous success of the commercial innovation and transformation processes: the acquisition of market share

In spite of the revolutions that we observed in the period taken into consideration, the contextual transformation of the production and commercial parts of the company have created important virtuous synergies, which in turn have given excellent results in terms of market share recovery.

Such growth in the market is another proof of the company's continuous success. As we already mentioned in the second part of this book,[2] TMHIT market share has grown considerably, despite the game-changing transformation of the sector and of the reference markets: the 17.5 per cent share recorded in 2009 has become 23.44 per cent in 2015. A regular, steady improvement ensured the success in all the sectors that form the Italian market for forklifts:

- the market for forklifts destined for logistics organizations and large retailers grew from 17.4 per cent in 2009 to 23.22 per cent in 2015;
- the market for electric frontal forklifts saw an increase from 20.7 per cent to 31.96 per cent over the same period;
- the market share for diesel and gas-powered frontal forklifts also recorded a larger market share, shifting from 10.4 per cent to 11.88 per cent.

7. Development and extension of corporate transformation

The real technology, force and energy behind our corporate transformation was represented by people, by their determination, motivations, incentives and professional qualifications.

People, their essence, their operational professionalism: that is what constitutes the propulsive core, the real force behind the platform of Toyota principles and values. Their knowledge and competence have been, and will continue to be, at the heart of each management initiative, focusing on people as the real engine that ensures continuity and innovation during our relationship with the market and the final user.

The growth and consistency of the market success achieved by the Italian Toyota Material Handling organizations were determined by the contaminating extension of the methods applied, by the operational interpretation and transformation of Toyota's principles into winning choices and virtuous practices, implemented by management towards external organizations and employees.

In particular, the relationship based on trust between company and dealerships represents the 'beating heart' of the development of business opportunities into sales and additional market presence.

An increase in knowledge by the network of dealers, supported by the company's management, integrated the joint development of traditional and innovative services and problem-solving techniques, creating added value and tailor-made support for customers. Such activities were fundamental for retaining existing customers and maintaining the company's share in the reference market.

Finally, the continuous innovation of the company offerings, developed and enhanced with the addition of new educational services, was a further driver for differentiation which contributed to consolidate interest and reach new customers. The innovation of consultancy services on offer reinforced the expansion of the processes aimed at customer retention, helping to turn opportunities into tangible business relationships; it also strengthens the commercial partnerships that in turn bring new cooperation advantages.

8. Development and extension of the continuous and transformative success outside the company, towards the business community, with the initiatives of Toyota Academy

The offerings differentiation and the hunt for new business opportunities have been supported by a dedicated organization, set up and developed to perform the following activities:

- to support the dealerships during the early stages of this reorganization, developing and fine-tuning the skills and knowledge of those employed in sales, with innovative training programmes, consultancy and support services;
- to extend services, consultancy and training activities to other businesses and organizations belonging to various business communities.

Toyota Academy was created with the goal of designing, developing, promoting and offering new managerial and technical consultancy services for production, operations, strategies, organization and logistics. We assembled a team of expert managers and professionals with a solid and highly specialized technical knowledge background acquired inside Toyota Material Handling, where they held positions of responsibility in several business departments (such as research and development, operations management, factory organization): experts in technologies and factory production processes, experts in logistics and advanced logistics, experts in business and commercial organization, and visual and business management consultants.

Over the years, Toyota Academy has developed several consultancy projects, requested by companies belonging to different business communities in Italy, projects that consistently achieved remarkable success.

Toyota Academy has been organizing technical and specialized workshops for years, held directly inside the plant, while every year it holds an exclusive 'Development Tour for High Technical Experience', which includes direct access to the Toyota factories in Japan.

Toyota Academy takes part in a number of university courses and master's degrees with senior lecturers, and it offers in-house training projects to Italian entrepreneurs, managers, professionals, businesses and organizations, aiming to spread the successful best practices applied by Toyota.

9. Development and extension of the continuous and transformative success outside the company: the transformation of the network of relationships in the territory and with other stakeholders (schools, universities, etc)

Witnessing the level of continuous success that the company achieved, the management and employees have been awarded several prizes by the European organizations of Toyota, in recognition of the amazing results that were posted.

The awards were given to Toyota Material Handling in Italy for reaching significant goals and targets, including:

- performance and excellent commercial results achieved (market share increase, sales targets hit);
- development of the organization and the levels of excellence reached by the plant;
- unquestionable success transforming the company and reaching the quality and excellence targets through continuous improvement of the processes;

- results obtained in quality improvement;
- recognition by the parent company for our excellence in *Kaizen* activities.

Toyota Material Handling has promoted its presence in Italy with a regular interaction and collaboration with schools and universities in the area as well as in other regions:

- following requests from faculties, institutes and business schools, a large number of master's degree theses, studies, projects and elaborations for master's and post-university specialization courses;
- organizing guided visits to the factory dedicated to students;
- participating in public educational events of institutional and social relevance;
- sharing the managerial and organizational experience and the success of the company with organizations and private and public institutions during events organized by the stakeholders and at social public events and meetings;
- explaining to university students the concepts and principles of Lean leadership and TPS, illustrating the advantages of implementing the differential operational approach of the Toyota Way principles;
- organizing several guided visits to the factory and training workshops for businesses, manager teams, entrepreneurs, consultants, and for all those who wish to understand the managerial systems and the successful practices applied by Toyota Material Handling in Italy.

Continuous success and the value-generating wave of the new cooperative management 4.0

The continuous and systemic success achieved by the Italian Toyota Material Handling organizations was the result of the choices made and of the challenges won by management, which has first and foremost focused on the differential application of the values of the Toyota Way.

The Italian management of Toyota Material Handling saw people at the strategic core of the new approach, giving total priority to teamwork, developing a structured and appropriate cooperative activity. Although those structured cooperative behaviours do not always mean excellent results, management still succeeded in creating a satisfactory level of cooperation, which supported continuous success and resulted in a value-generating wave of the new cooperative management 4.0.

In particular, the Italian management of Toyota Material Handling:

a learned and acquired the principles of Lean leadership, Lean organization and total quality;

b absorbed and metabolized the values of the Toyota Way;

c studied the methods applied by the Japanese management to TPS, the universally known Toyota Production System. The management also introduced technical and operational modifications, continuously improving the successful practices implemented by Toyota's Japanese plants, which were visited several times by the Italian management of Toyota Material Handling Italia.

During said visits, the management had the opportunity to focus and exchange opinions on the issues that arose in Italy, refining the techniques and the methods to apply them and to define the strategic approaches that had great success in Toyota's factories and divisions.

Moreover:

- Toyota's Italian management has decided to face the challenge posed by the creation of a common teamwork attitude, introducing and developing an indispensable shift in the company's approach, leaving the previous corporate culture that, just as with most of the Italian companies, gave priority to individual rather than collective performance.
- During its effort to achieve the required cultural change, it underlined the need for continuous exchange and sharing of knowledge, capabilities, experiences and results achieved, be them in the factory or in other sectors of the company. It succeeded by taking advantage of several visits made by the Japanese 'super experts', learning and communicating in a constructive manner. Those visits were the occasion to focus and exchange ideas about new techniques and methods of implementation of TPS and the Toyota Way, and also to discuss the operational methods for production and management.
- It reconfigured and adapted the principles of TPS and the Toyota Way values, fitting them into specific scenarios typical of Italian organizations and factories, with limitations, cultural hurdles and notable differences in organization, production and sales.

- It did all it could to eliminate excuses and pretexts, to overcome reluctance and resistance, to create optimism and belief in the adoption and implementation of the new techniques, methods and innovative processes.

The same cultural shift achieved the following results:

- It defeated the pockets of resistance that opposed the change, the rooted and outdated philosophy of 'autonomous and isolated factory worker' and the reluctance to share, training employees and convincing them of the importance of cooperation, teamwork, sharing mistakes and improvements.
- It succeeded in the difficult job of transforming the culture of the company, overcoming the prejudices deriving from a cultural resistance to teamwork and from the habit of a consolidated process: 'We have worked like this for years; why should we change?' was one of the most reported objections, which ascertained the discomfort and the initial widespread resistance to change.

Slowly, the persistence of management and of its nearest collaborators started to get results and spread the idea of sharing, of teamwork energy, of the will to improve continuously.

It did not happen overnight, but the cooperative approach eventually managed to overcome the old, divisive and patchy culture: the importance of teamwork, of the techniques for sharing, of the collaboration required to achieve success and of the new good practice of continuous improvement.

The removal of errors and waste and the constant focus towards improvement are now the elements that determine the success of the company and drive the new corporate culture. Changing behaviour and moving towards continuous improvement and sharing have become part of the new company's DNA.

In short, the new culture has defeated the old behaviours and customs, creating a new climate of continuous sharing and communication and a new victorious, structured constructive collaboration: cooperative management has succeeded.

The management of Toyota, by involving and motivating the workforce, has also managed to:

- change its management and command style, developing a new multipolar, inclusive and constructive leadership model. Such leadership prevailed also with the valorization and sharing of small and fast operative results ('quick win') that the work teams achieved;

- establish an environment where optimism and trust are common, facilitating the contribution of every single person even after the implementation and testing of everyday change. The development and affirmation of the new group spirit has promoted the adoption of new interaction methods, of common knowledge and experiences, of constructive communication, which has sped up the process of corporate transformation;
- speed up the introduction of visual management in every production environment (factory and production departments), boosting sharing, discussion and implementation of continuous improvement processes.

Other achievements reached by the Italian management of Toyota include:

- the creation of its own unique business vision and commercial action, refining its capability to innovate, create differential value and remain one step ahead of competitors in terms of meeting the customers' needs;
- the gradual extension of the new strategic visions and methods, successfully explaining to everyone both inside and outside the company that gradual improvements and innovation leaps can be compatible, sometimes crucially so.

In short, the extension of the **new cooperative and transformative management 4.0** set up and developed by management, a new culture of change and continuous improvement, contributed to strengthen results and performance, in particular where it was possible to create symmetries, strategic and operative synergies with suppliers and customers.

The adoption of the new cooperative and transformative management 4.0 allows a **new virtuous transformation circle** to be developed, an ecosystem that creates a set of value innovative chains, integrating downstream and upstream organizations to create new networks, chains, constellations and ecosystems capable of amplifying the results obtained.

Management also succeeded to create a constant communication flow, an endless transmission and exchange of knowledge and experience that now come naturally. With teamwork and cooperative management, it managed to develop new collaborative methods that were critical to overcome issues and solve problems.

We managed to reduce complexity through work and cooperation, overcoming the communication barriers among people, solving stalled situations and operative gridlocks both inside and outside the factory.

The development of commercialization and go to market processes brought the right level of interaction, relying on the additional value of the knowledge and experience of other people: company workers, external entities (dealerships, other chain operators) and operators working for other companies, such as with the consultancy activities developed by Toyota Academy.

These changes, transformations and value created have not only contributed to the company's continuous success, but also to the constitution of a qualified value-generating wave of the new cooperative management 4.0 and to the redefinition of the DNA of the Italian organizations of Toyota Material Handling, of their cooperative core and of their search for constant innovation.

The continuous success of the Italian Toyota Material Handling organizations driven by the company's commercial sector

The continuous success of the Italian organizations of Toyota Material Handling is due to the requests for innovation of products and services by the commercial area of the plant, alongside the need to align and conform the range and the production methods to the evolution of customers' needs.

The main driver for transformation is represented by customer needs. The commercial organization alongside marketing and sales are the main figures of the entire business transformation. In other words, customer needs and requests from TMHIT's commercial area are the foundations for the transformation of the factory.

The changes of customer requirements and market evolutions were the real drive behind strategic and managerial business transformations for years, and they will be even more influential in the future; company managers can see this happening daily, with increasing relevance. Logistics and warehouse needs are less standardized and more specific, harder to manage properly.

Every company is specializing in logistics, each one following its own logic. An example is given by Ikea, where the demand for products has specifications that are increasingly customized, far from standardization.

The demand for standard forklifts has been replaced by requests that are specific, dedicated, custom-made. If the producer does not adjust quickly to personalized, non-standard products, with different specifications and final use, the risk is to have an offer that does not meet market demands, eventually losing customers that the company succeeded to retain for years.

The introduction and global adoption of automated warehouses helps the creation of non-standard and customized unique solutions. Each company works to be aligned with the market; in time, standard products will lose their appeal and market share. The decrease of orders for standard forklifts is also caused by the change in business logistics solutions. Years ago, demand was based on standard Euro pallett measuring. Nowadays, things have drastically changed and require continuous adjustments, making production more difficult. Not responding to the needs of customers means that the company is far from meeting the new requirements for logistics efficiency. This obviously causes the deterioration of the relationship with the customer, which will take its business to a company that can adapt its products accordingly.

As we mentioned previously, continuous success was the result of the challenges that the management of Toyota Material Handling in Italy faced and won during the period assessed in this book (2000–16), when the results have been solid, consistent and repeated.

Such success proves the reliability of the choices made and of the methods adopted by the Italian management of Toyota Material Handling.

The constant success in terms of results over the long term by the Italian organizations of Toyota Material Handling show how this case history is an example of managerial excellence, which needs to be studied and focused upon, not only for the quality and quantity of the results, but also for their steady volumes even during a period of deep crisis.

The results obtained by Toyota Material Handling in Italy can be ascribed to the system of challenges taken and won by management, to the right application and reinterpretation of the platform, principles and values of Toyota.

Notes

1 Intended as a systemic set of different innovation steps of various processes, initiatives and transformation activities, of complex change processes planned and enacted by the management of the Italian organizations of Toyota Material Handling.

2 See Figures A.7a and A.7b of Appendix A.

GLOSSARY

4M – 4M represents the diagram developed by Kaoru Ishikawa, Tokyo University professor, one of the main promoters of total quality. 4M summarizes the four main elements that affect the industrial process:

- Man. It includes all activities linked to human resources, such as manual production activities, planning, etc.
- Machine. It takes into consideration all activities related to machines. It also includes equipment, work tools, energy, etc.
- Method. It includes all activities linked to work methods, such as instructions, procedures and flows.
- Materials. It includes all activities related to the materials that are part of the process, such as raw materials and semi-finished products.

4M is split into sub-categories until the main causes for an issue are detected. This method is also called fishbone or cause–effect diagram. Some versions include factors such as measuring or equipment in relation to the processes under analysis.

5S – The initials of five Japanese words that summarize the five steps required for their implementation. They are at the basis of any development activity undertaken following TPS. Aside from being used to keep a clean and organized environment, they have a deeper meaning because they drive the search for anomalies (the concept of *Jidoka*). Following the principle 'a place for everything and everything in its place', the working area becomes an environment where each item has its specific place. This also applies to the balancing buffers between processes: since they are dimensioned to hold a clear, predefined number of parts, any anomaly is immediately spotted.

- *Seiri* (Separate). Separate what is needed from what is not needed.
- *Seiton* (Set in order). Arrange needed items in designated locations and visually mark where they belong.
- *Seiso* (Shine). Clean any kind of dirt in one's working area.
- *Seiketsu* (Standardize). Standardize by defining a repetitive and continuous system.
- *Shitsuke* (Sustain). Spread, share, support, respect and maintain.

In Japan it is called '4S + Discipline', since the last S, the most complex of them all, implies a deep involvement of human resources and management to maintain the level reached with the previous 4S.

Andon – The *Andon* is a lamp made with a bamboo, wood or metal frame, over which a rice paper sheet is laid to protect the flame from the wind. Toyota relied on this concept to develop its signalling system, which highlights anomalies along the production line using a simple light.

Each workstation is equipped with two ropes (or buttons), yellow and red; in case of problems, the operator can call the TL with the yellow rope. If the problem is not solved within the allotted takt time, using the red rope stops the production line, which is one of the pillars of TPS, the *Jidoka*.

The *Andon*, installed in all Toyota plants, shows the main information related to production, such as the daily target, the progressive target and the production volume reached, the hours of production planned for the day (including overtime) and the period of downtime.

Asaichi – Literally 'morning market'. It's a morning meeting to discuss and visualize the problems encountered the previous day. It is managed by the group leaders or by the operators, who explain the problem with the aid of images or with the affected part itself.

It is extremely important that:

- the meeting area must be as near as possible to the reference area (production, offices, etc);
- all issues must be visualized and visible to everyone;
- KPIs, trends and targets must be indicated;
- each report must be managed with the PDCA and must indicate the cause, the countermeasure, the person in charge of its solution and the deadline.

Dantotsu – This Japanese word can be translated as 'the best of the best'. It was used by Toyota Material Handling Group for a quality improvement project inside the plants. The message that it conveys is: 'to unquestionably be the best for the quality of our products'.

Dojo – Literally, it means 'place (jo) where the way is followed (do)'. In the Western world, the term is wrongly translated as 'gym' and it is used only to describe an area for training. The oriental culture sees *dojo* as the place where you can follow the way and reach perfect communion between *zen* (mind) and *ken* (body), achieving perfect balance and maximum expression of one's individuality. *Dojo* is the school of the *sensei* (master):

he is the apex, his are the guidelines and rules for its good management. The *sensei* is supported by other teachers, by his former pupils and by the *senpai* (senior students), who play an important role: their daily behaviour is the example that other apprentices must follow. When a *senpai* does not care about his behaviour, he becomes a liability for the whole school. At Toyota, *dojos* are places dedicated to training; specifically, TMHMI uses a *safety dojo* and a *training dojo* to train operators.

Genchi Genbutsu – One of the values of the Toyota Way. It remembers the importance to 'go to the source, where things happen'. It is one of the key concepts of the Toyota Production System, highlighting the importance of going and visiting personally, talking to those who are actively involved on the line. This concept is taken into great consideration by Toyota's main management, which periodically visits the factories and the sales offices around the world.

Heijunka – Technically, it means 'to level production'; it is one of the principles of the Toyota Production System. An unbalanced production causes peak production times that might overwork the operators, raising the risk of injuries and quality issues. While planning the production sequence, it is important to take this factor into consideration, alternating labour-heavy products with those with lighter involvement by the worker.

Jidoka – Together with Just In Time, it is one of the pillars of the TPS. Its origins lay with the founder of the company, Sakichi Toyoda, who invented a loom that could stop automatically if one of the threads broke. This principle is still followed today: inside the Toyota production lines, if a problem is not solved within the takt time, the line is stopped. One of the main concepts of the TPS is, in fact, to 'bring anomalies to light'.

Jishuken – The term translates as 'independent managerial study groups'. The root philosophy of the groups is the *Kaizen*, but in this case the managers are those who identify the elements that need to be improved, and they are personally part of the work group.

Jundate – This term identifies the procedure of giving a logical sequence to the components (in particular to the bulkier items), so that they are placed as near as possible to the assembly line and reduce the waste caused by operator movements. TMHIT organized areas where the parts are picked according to the sequence along the assembly line, transported or positioned next to their area of destination with small trains.

Junjo – Literally, it means 'succession, sequence'. It identifies the production sequence (and the products) frozen by planning: the progressive number assigned becomes the unique reference point for all the company activities. The sequence is transmitted to the supply chain, which delivers the materials based on the number of agreed sequences. Internal logistics use *Junjo* to identify each part or kit involved in the production cycle.

Just In Time – One of the pillars of TPS. It was created by the founder's son, Kiichiro Toyoda, who implemented it in the new company that he founded to enter the automotive business. Just In Time can be summarized with a single principle: 'what is needed, when it is needed and where it is needed'. Its fundamental concept lays in the switch of production logic from Push to Pull. In practical terms, it is a policy of stock management that uses methods that can improve the production process, optimizing the upstream stages, minimizing the stock of raw materials and components needed for production. It implies coordinating the times when the parts are physically needed on the line with their purchase and availability.

Kaikaku – Japanese for 'radical change'. Usually it is an activity driven by top management that implies a substantial change of technology, know-how or strategy. It implies the introduction of new technologies or equipment of a certain importance, which can bring great improvements to production performance.

Kaizen – The term is formed by *kai* and *zen*, two ideograms (改善) that mean, respectively, 'change' and 'better'. It is the continuous improvement achieved by small steps, in juxtaposition with radical change, which is indicated with the term *ku*. Toyota pays a lot of attention to the *Kaizens* suggested by direct operators, promoting meetings, international gatherings (*Azora Ichiba*) and awards (usually not in cash) assigned by the group's president for the most brilliant *Kaizen*.

Kaizen Kobo – Area for the production of small tools, usually related to the implementation of *Kaizen* ideas. It includes pipes, belts and joints that allow a wide range of tools to be built quite easily, adapting them to the specific needs of the operator.

Kanban – Literally, it means 'card, sign'; it's a system developed by Toyota to manage the restocking of materials. It is a typical Pull restocking model, because it is the actual consumption, rather than the forecast, to activate the reordering of materials. After the first (or the last) item in the container is retrieved, the operator takes the *Kanban* card and places it in a collection centre (*Kanban* board). The card is the signal that activates restocking, which can be external (the information is relayed to the supply chain) or internal (the parts are retrieved from the supermarket).

Mono* and *Hito Zukuri – They represent the symbiosis between people and production. The term *Hito* means 'people'; the same applies to *Mono*, but the latter also means 'things, quality product'. The terms put together define the concept of 'capability of making quality products with people'.

Mieruka – It is an important concept of TPS: visualization. There are two different kinds of *mieruka*: the first (視える化) refers to watch, observe, distinguish, study, while the second (見える化) is limited to a superficial look. Visual management is very demanding and must activate the cycle of *Kaizen*, or continuous improvement (Manisera, 2010).

Muda – A Japanese term that defines the activities without added value for the customer. In Toyota there are seven different types of *Muda*: transportation, inventory, motion, waiting, over-processing, over-production and defects. The last *Muda* is the worst of them all, because it spawns all the others. In addition to *Muda*, Toyota has defined two other kinds of waste: *Mura* and *Muri*. *Mura* is the waste caused by production peaks or by unbalanced processes. *Muri* is the waste caused by overstressed people, due for example to excessive workload, insufficient training or lack of information.

Obeya – Literally, 'big room' or 'war room'; it is generally used by Toyota project managers to visualize the activities and the processes related to a specific project. It was first used during the creation of the hybrid vehicle Prius, because it was necessary to put together knowledge and skills of various company segments. The organization of meetings inside the *Obeya*, where the walls displayed all the relevant development activities, was the ideal method to manage the work done around this important model.

QCC – Quality control circles, which are systematically carried out inside Toyota. They involve all the employees, are composed of groups of about 10 people, coordinated by the team leaders and held in the production areas. In order to implement the circles, the participants are trained in problem-solving definitions and methods. All production activities are stopped once or twice every month in order to carry out the QCCs.

Yamazumi – The *Yamazumi* chart is an analysis model of cycle times that splits the activities on each workstation into value-adding and non-value-adding. Using a graph where the x-axis shows the stations and the y-axis marks the times, a histogram displays the times for each station, highlighting value-adding activities and separating them from those that do not generate value for the customer. By tracing the takt time on the graph, you can detect the level of saturation of each workstation and proceed with further balancing of their activities.

YK-Y – The acronym for *Yarinikui* (operations difficult to perform), *Kizukai* (operations hard to perform) and *Yarinahoshi* (operations to repeat, correct); a tool that in conjunction with the QCC can be used directly by employees to suggest solutions that can improve the workstation's performance. It is coordinated by the TL, who collects the ideas coming from the team and presents them to an inter-function committee.

BIBLIOGRAPHY

Alberti G.E., Gandolfi A., *La pratica del Problem Solving*, Franco Angeli, Milano 2008.

Amplatz C., *La comunicazione in aula. Introduzione alla problematica*, Tecomproject Ed. Multimediale, Ferrara 1998, pp. 74–77.

Chiappi R., *Problem Solving nelle organizzazioni: idee, metodi e strumenti da Mosè a Mintzberg*, Springer, Milano 2006.

Ciappei C., Citti P., Bacci N., Campatelli G., *La metodologia Sei Sigma nei servizi, Un'applicazione ai modelli di gestione finanziaria*, Firenze University Press, Firenze 2006.

Daft R.L., *Organizzazione aziendale*, edizione italiana a cura di Boldizzoni D., Apogeo editore, Milano 2017.

D'Amore B., *Problemi*, Franco Angeli, Milano 1996, p. 298.

Dunker K., *On problem-solving. Psychology Monographs*, 58, 1945 (It. trans. *La psicologia del pensiero produttivo*, Giunti-Barbera, Firenze 1969).

Fantin I., *Applicare il Problem Solving*, CreateSpace Independent Publishing Platform, 2013.

Fischetti A., *La Creatività e il Problem Solving*, Alpha Test, Milano 2009.

Galgano A., *I sette strumenti della qualità totale. Manuale operativo*, Il Sole24Ore, Milano 2001.

Kanizsa G., 'Il "problem solving" nella psicologia della Gestalt', in Mosconi G., D'Urso V. (a cura di), *La soluzione dei problemi*, Giunti-Barbera, Firenze 1973, p. 35.

Koomey J.G., *I numeri che contano. E l'arte del problem solving*, Muzzio, Padova 2009.

Laeng M., 'Enciclopedia pedagogica', La Scuola, p. 9397.

Liker J.K., *The Toyota Way*, McGraw-Hill Education, New York 2004.

Montgomery D.C., *Controllo statistico della qualità*, McGraw-Hill, Milano 2000.

Mosconi G., D'Urso V., *La soluzione dei problemi*, Giunti-Barbera, Firenze 1973.

Nardone M.C., *La terapia dell'azienda malata*, Ponte alle Grazie, Firenze 2000.

Nonaka I., Takeuchi H., *The knowledge creating company*, Oxford University Press Inc., Oxford 1995.

Ohno T., *Toyota Production System: Beyond Large-Scale Production*, Productivity Press, New York 1988.

Pande P., Holpp L., *Sei Sigma, Guida introduttiva*, edizione italiana a cura di Punzo S., Debernardi M., Colella M., Tecniche Nuove, Milano 2005.

Rogers E.M., *Diffusion of Innovations*, Free Press, New York 1983.

Schein E.H., 'Coming to a New Awareness of Organizational Culture', *Sloan Management Review*, Winter 1984, pp. 3–16.

Shigeo Shingo, A.P., *A Revolution in Manufacturing: The SMED System*, Productivity Press, New York 1985.

Smalley D.K., *Understand A3 Thinking*, Productivity Press, New York 2012.

Trevisani D., *Regie di cambiamento*, Franco Angeli, Milano 2007.

Verga D., Munari L., *La qualità dei servizi bancari per le piccole e medie imprese*, L'impresa Banca, Credito Italiano, settembre 1987.

Verga D., *Fattori di successo e strategie delle piccole e medie imprese italiane*, Istituto Universitario di Bergamo 1988.

Verga D., *L'evoluzione dell'impresa commerciale: Aspetti teorici, condotte di mercato e strategie competitive*, Libreria Nazionale, Roma 1990.

Verga D., 'Advanced Business Management: il nuovo modello per migliorare la profittabilità, per affrontare con successo e comunicare efficacemente sui mercati internazionali', in *Mercato Globale*, Milano, febbraio 1997.

Verga D., 'Comunicazione e Marketing aperto: nuovo management e percorsi operativi per l'attivazione della Comunicazione e promozione nei mercati internazionali', in *Mercato Globale*, Milano, 2 aprile 1997, pp. 8 e seguenti.

Verga D., *Plurimarketing bancario: Lo sviluppo delle competenze di Marketing, Competenze per competere*, Franco Angeli, Milano 1999.

Verga D., Grasso M., *Glossario di Marketing e comunicazione. Le parole della next economy, del Net Marketing e della Web Communication che ogni manager deve conoscere*, Franco Angeli, Milano 2001.

Verga D., *Competenze Digitali, L'impresa flessibile*, Franco Angeli, Milano 2004.

Verga D., *Net-crazia: le best practice in rete*, in 'Innovazione e creatività nell'era digitale', Lamborghini B. e Donadel S. (a cura di), *Le nuove opportunità della digital sharing economy*, Franco Angeli, Milano 2006.

Verga D., 'Small Business Management: il nuovo management responsabile "infinito punto zero". La crescita delle piccole e medie imprese e lo sviluppo delle performance aziendali nei periodi di crisi' in *L'impresa web. Social Networks e Business Collaboration per il rilancio dello sviluppo*, testo a cura di Lamborghini B., Franco Angeli, Milano 2009.

Von Krogh, Ichijio K., Nonaka I., *Enabling knowledge creation*, Oxford University Press, Oxford 2000.

Womack J.P., *The Machine That Changed the World: The Story of Lean Production*, Harper Business, New York 1991.

ACKNOWLEDGEMENTS

Heartfelt thanks to:

- Ambrogio Bollini, for his role in guiding the company during these years, making it grow and become a reference point within TMHE.
- TMHMI and TMHIT management, responsible for the results that were achieved by the companies.
- The members of TMHMI Steering Operations, exceptional leaders who interpreted and transmitted the values of Toyota and TPS, motivating and guiding their team.
- All the group leaders, team leaders and managers who actively contributed to this transformation.
- All the Japanese senior advisors of Toyota, who mentored, guided and supported us during our transformation.
- TMHMI and TMHIT supply chain, which contributed to our success.
- All the TMHIT agents and dealers for their trust and cooperation during their activities. A significant part of our continuous success was achieved thanks to their passion and commitment.
- Carpenfer S.p.A. in the person of Dr Sandro Bordoni; Enersys Italia S.r.l. in the person of Dr Maurizio Dal Sasso; Bonfiglioli S.p.A. in the person of Sonia Bonfiglioli, engineer; Würth Italia S.r.l. in the person of Nicola Piazza, engineer. All of them helped with their contributions to the creation of this book.

INDEX

Page numbers in *italic* refer to photographs or tables.